Cosmic Odyssey:

Exploring the Universe and Unveiling the Alien Enigma

By Brandon Taul

Contents

Chapter 1: Prologue to the Cosmos 4

Chapter 2: Our Cosmic Neighborhood 14

Chapter 3: Galactic Splendor 26

Chapter 4: The Great Beyond: Intergalactic Marvels 38

Chapter 5: Alien Worlds: A Search for Life............................49

Chapter 6: UFO Phenomenon: Fact or Fiction? 61

Chapter 7: Communication with Extraterrestrial Intelligence...75

Chapter 8: First Contact: Fictional Scenarios 88

Chapter 9: The Fermi Paradox: Where Are They?.................98

Chapter 10: The Future of Space Exploration and Alien Contact 106

Chapter 11: Epilogue: Embracing the Unknown....................117

Appendix: Resources and References 127

This book

will blend scientific

facts with speculative fiction

elements, taking readers on a

journey through the known universe

while delving into the fascinating concept

of extraterrestrial life. It will encourage

readers to ponder the mysteries of

the cosmos and embrace

the possibility of

connecting with

intelligent beings

from beyond our world

Chapter 1: Prologue to the Cosmos

Introducing the vastness and mysteries of the universe

The Cosmic Canvas

In the vast expanse of the universe, countless galaxies, each containing billions of stars, dance across the cosmic canvas. The observable universe, spanning nearly 93 billion light-years in diameter, evokes a sense of awe and wonder in the hearts of astronomers and stargazers alike. To put this staggering scale into perspective, consider that a beam of light, which travels at an astonishing 186,282 miles per second, would take nearly 93 billion years to traverse the entire breadth of the observable universe. Indeed, the cosmos is a vast and mysterious place, inviting us to embark on a journey of discovery. The universe we see today is a product of an event that defies our conventional understanding of space and time - the Big Bang. Approximately 13.8 billion years ago, the universe sprang forth from an unimaginably hot and dense point known as a singularity. In a fraction of a second, the cosmos began expanding at an exponential rate, giving birth to space, time, and the fundamental forces that govern the cosmos. As the universe expanded and cooled, matter and energy coalesced, eventually forming the galaxies, stars, and planets we observe today.

The Birth of the Universe

The early universe was a seething cauldron of particles and radiation. As it expanded and cooled, subatomic particles started to combine, forming the first atoms of hydrogen and helium. These atoms clumped together under the influence of gravity, giving birth to the first structures in the cosmos - vast clouds of gas and dust known as nebulae. Over millions of years, gravitational forces caused these nebulae to condense further, leading to the formation of the very first stars. In the hearts of these nascent stars, nuclear fusion ignited, converting hydrogen into helium, and releasing an incredible amount of energy in the process. The universe had witnessed the birth of its first stellar inhabitants. These early stars were giants, much larger and hotter than most stars we see today. They lived fast and died young, culminating their brief but brilliant lives in spectacular supernova explosions. The remnants of these ancient stars now enrich the cosmos with the heavy elements necessary for the formation of planets, moons, and life as we know it.

Cosmic Evolution

As eons passed, the universe underwent a remarkable evolution. Within the vastness of space, galaxies emerged and evolved into diverse shapes and sizes. Spiraling galaxies like the Milky Way, with their graceful arms adorned by countless stars, twirled in cosmic dance. Elliptical galaxies, resembling large, glowing orbs, hosted ancient populations of stars. Irregular galaxies, with their chaotic and asymmetrical appearances, defied traditional classifications. Within these galaxies, stars continued to be born and die, shaping the destiny of the cosmos. Stellar nurseries, dense regions of gas and dust, gave birth to new generations of stars. These celestial infants, often enshrouded by their natal clouds, eventually emerged to illuminate the cosmos with their brilliance. However, the life of a star is finite, and each stellar journey concludes in a grand finale. Massive stars, upon exhausting their nuclear fuel, undergo cataclysmic supernova explosions, releasing vast amounts of energy and scattering heavy elements into space. The remnants of these supernovae form exotic objects like neutron stars and black holes, the enigmatic cosmic monsters with gravitational pulls so intense that not even light can escape their clutches. But from the ashes of stellar demise emerges new hope for celestial life. The enriched interstellar medium, now seeded with heavy elements, serves as the fertile ground for the formation of future generations of stars and planetary systems.

Galactic Splendor

At the heart of our cosmic journey lies the Milky Way galaxy, our home amidst the vast sea of stars. This grand spiral galaxy, with its luminous band of stars stretching across the night sky, has captured human imagination for millennia. Our solar system, located in one of the spiral arms of the Milky Way, orbits the galactic center approximately once every 230 million years. Surrounding us are numerous neighboring galaxies, each with its own tale to tell. The Andromeda Galaxy, a magnificent spiral akin to our own Milky Way, lies at a distance of about 2.5 million light-years away, making it the closest major galaxy to our own. The Large and Small Magellanic Clouds, two irregular dwarf galaxies, dance in a gravitational tango with the Milky Way, their presence a testament to the interconnectedness of the cosmos.

Beyond our galactic neighborhood, a cornucopia of galaxies awaits exploration. Clusters of galaxies, bound by gravity, form gargantuan superstructures in space. The Virgo Cluster, Coma Cluster, and Perseus-Pisces Supercluster are among the cosmic cities of galactic residents. Yet, amidst the splendor of the galaxies, space also reveals vast voids where galaxies seem conspicuously absent. These cosmic voids, akin to celestial deserts, challenge our understanding of the universe's structure and expansion.

Historical Milestones in Space Exploration

The human fascination with the cosmos and the desire to explore beyond our world have driven us to embark on an incredible journey of space exploration. Over the course of the last century, we have achieved remarkable milestones that have not only expanded our knowledge of the universe but also revolutionized our understanding of ourselves and our place in the cosmos.

The Space Age Dawns

The Launch of the First Artificial Satellite: Sputnik 1 (1957)

On October 4, 1957, the Soviet Union launched Sputnik 1, the world's first artificial satellite, into Earth's orbit. This historic event marked the beginning of the space age and triggered the space race between the United States and the Soviet Union during the Cold War era.

Yuri Gagarin's Historic Spaceflight (1961)

On April 12, 1961, Soviet cosmonaut Yuri Gagarin became the first human to journey into space aboard the Vostok 1 spacecraft. His orbit around Earth lasted just 108 minutes, but it was a giant leap for humankind and a testament to the progress of space exploration.

Pioneering Lunar Expeditions

Apollo 11: Humanity Sets Foot on the Moon (1969)

On July 20, 1969, NASA's Apollo 11 mission achieved the pinnacle of human exploration by landing astronauts Neil

Armstrong and Edwin "Buzz" Aldrin on the lunar surface. Armstrong's iconic words, "That's one small step for man, one giant leap for mankind," marked a defining moment in history.

The Moon Reveals Its Secrets (Late 1960s - 1970s)

Through subsequent Apollo missions, astronauts brought back invaluable lunar samples and conducted experiments, shedding light on the Moon's geological history and its role in the early solar system.

Space Probes and Planetary Exploration

Mariner and Viking Missions (1960s - 1970s)

NASA's Mariner missions, including Mariner 2, provided humanity's first close-up observations of Venus, Mars, and Mercury. The Viking missions, which landed on Mars in 1976, became the first successful landings on another planet.

Voyager 1 and 2: Voyages to the Outer Solar System (1977)

Launched in 1977, Voyager 1 and Voyager 2 have become legendary ambassadors of Earth. They provided stunning images and crucial data about the outer planets, their moons, and the heliosphere's boundary. Voyager 1 is now the farthest human-made object from Earth, venturing into interstellar space.

Space Shuttle Era

Space Shuttle Program (1981 - 2011)

The Space Shuttle program marked a new era in human spaceflight, with reusable spacecraft capable of carrying astronauts and payloads into orbit. The Shuttle missions played a vital role in building the International Space Station (ISS) and conducting numerous scientific experiments in microgravity.

International Collaboration and Space Stations

International Space Station (1998 - Present)

Launched in 1998, the ISS represents a multinational effort, involving space agencies from the United States, Russia, Europe, Japan, and Canada. The ISS serves as a research laboratory in low Earth orbit, conducting experiments that benefit life on Earth and prepare for future deep space missions.

Robotic Exploration and Beyond

Mars Rovers (2000s - Present)

NASA's Mars rovers, including Spirit, Opportunity, and Curiosity, have significantly advanced our understanding of the Red Planet. These robotic explorers have traversed the Martian surface, analyzing rocks, soil, and searching for signs of past or present life.

New Horizons: A Glimpse of the Distant Solar System (2015)

In 2015, NASA's New Horizons spacecraft performed a historic flyby of Pluto, revealing a world of icy plains and towering mountains. The mission provided the first up-close images and data of this distant and enigmatic world.

The Quest Continues

From the first steps in space to the distant edges of the solar system and beyond, humanity's journey of space exploration remains an ongoing adventure. The accomplishments and lessons learned from these historical milestones serve as stepping stones toward future endeavors, including crewed missions to Mars, space tourism, and the quest to discover life beyond Earth.

The Quest to Understand the Origins of the Universe

Since time immemorial, humans have gazed upon the night sky with wonder and pondered the origins of the cosmos. The quest to understand the birth of the universe, its underlying laws, and the fundamental forces that govern it has been one of the greatest intellectual pursuits of humanity. From ancient cosmological myths to modern scientific theories, our understanding of the universe's origins has evolved dramatically over the millennia.

Cosmological Myths and Early Theories

Mythology and Creation Stories

In ancient cultures across the globe, myths and creation stories were woven to explain the origin of the universe. From the Big Bang and cosmic inflation to the formation of galaxies and stars, these myths often depicted celestial beings shaping the cosmos and bringing life to the world.

Aristotle's Geocentric Model (4th century BCE)

The ancient Greek philosopher Aristotle proposed a geocentric model of the universe, with Earth at the center and celestial bodies orbiting it in concentric spheres. This model prevailed for centuries, shaping early cosmological thinking.

Copernican Revolution and Heliocentrism

Nicolaus Copernicus: Heliocentric Model (16th century)

In the 16th century, Nicolaus Copernicus revolutionized cosmology with his heliocentric model, placing the Sun at the center of the solar system. This groundbreaking idea challenged the geocentric view and laid the groundwork for modern astronomy.

The Birth of Modern Cosmology

Edwin Hubble's Observations (1920s)

In the early 20th century, astronomer Edwin Hubble made groundbreaking observations that revealed the expansion of the universe. He discovered that galaxies were moving away from each other, providing crucial evidence for the Big Bang theory.

The Big Bang Theory (1930s - 1940s)

Formulated by Georges Lemaître and further developed by George Gamow and others, the Big Bang theory proposed that the universe originated from a hot and dense state, expanding from an infinitely small singularity. This theory laid the foundation for modern cosmology and has since become the prevailing model for the universe's origins.

Cosmic Microwave Background and Inflation

Discovery of the Cosmic Microwave Background (1965)

Arno Penzias and Robert Wilson accidentally discovered the cosmic microwave background radiation, a faint glow of radiation permeating the universe. This discovery provided additional evidence for the Big Bang theory and supported the idea that the universe was once much hotter and denser.

Inflation Theory (1980s)

Inflation theory, proposed by physicists Alan Guth and Andrei Linde, suggests that the universe experienced a rapid expansion phase shortly after the Big Bang. This theory helps explain the observed uniformity of the cosmos and has become an integral part of modern cosmological models.

Observations and Precision Cosmology

Cosmic Structure Formation (1990s - Present)

Advancements in observational astronomy and precision cosmology have allowed scientists to study the large-scale structure of the universe. Sophisticated instruments like the Hubble Space Telescope and ground-based observatories have provided stunning images of distant galaxies and galaxy clusters, shedding light on the cosmos' evolving structure.

The Planck Satellite (2009 - 2013)

The European Space Agency's Planck satellite measured the cosmic microwave background with unprecedented accuracy. Its observations provided crucial data for understanding the age, composition, and evolution of the universe.

The Frontier of Cosmic Origins

Multiverse Hypothesis and Beyond

As our understanding of the universe's origins continues to evolve, new theories, such as the multiverse hypothesis, speculate about the existence of other universes beyond our own. These ideas challenge conventional thinking and inspire further exploration into the nature of reality.

The quest to understand the origins of the universe remains a captivating journey of human intellect and curiosity. With ongoing advancements in technology, observational capabilities, and theoretical frameworks, we are poised to uncover more profound insights into the universe's birth and the mysteries that lie beyond.

Chapter 2: Our Cosmic Neighborhood

Our Solar System: A Celestial Playground

The Sun: Our Life-Giving Star

At the heart of our solar system lies the Sun, a colossal ball of searing hot gas that provides light and warmth to our planet and its diverse inhabitants. The Sun's immense gravitational pull keeps the planets, moons, and other celestial bodies in orbit around it, shaping the delicate balance of our cosmic neighborhood.

The Inner Planets: Rocky Worlds

The innermost planets - Mercury, Venus, Earth, and Mars - are rocky worlds with solid surfaces. Mercury, the closest to the Sun, experiences extreme temperature fluctuations, while Venus is shrouded in a thick, toxic atmosphere, making it the hottest planet in our solar system. Earth, our home, teems with life and boasts a unique blend of geologic activity, weather patterns, and diverse ecosystems. Mars, often referred to as the "Red Planet," continues to captivate our imagination as we search for signs of past or present life on its barren surface.

The Gas Giants: Majestic and Enigmatic

Beyond the asteroid belt, the outer solar system hosts the gas giants - Jupiter, Saturn, Uranus, and Neptune. These colossal planets are primarily composed of hydrogen and helium, with thick atmospheres and a lack of solid surfaces. Jupiter, the largest of them all, plays the role of a cosmic guardian, protecting the inner planets from potential celestial threats by capturing or deflecting many incoming asteroids and comets.

Dwarf Planets and Asteroids: The Minor Players

Amidst the planets, we find a diverse array of celestial objects. Dwarf planets like Pluto and Ceres, once considered full-fledged planets, now inhabit a distinct category. The asteroid belt, situated between Mars and Jupiter, is home to countless rocky bodies of varying sizes and shapes. Some asteroids occasionally venture close to Earth, sparking both awe and concern among astronomers.

Moons: The Cosmic Companions

Earth's Moon: Our Lunar Neighbor

The Moon, Earth's only natural satellite, has been a source of fascination and inspiration for humanity throughout history. From its ancient impact craters to its changing phases, the Moon's presence in our night sky has influenced cultural myths, calendars, and scientific understanding.

Jovian Moons: A Menagerie of Worlds

Jupiter and Saturn boast numerous moons, some of which are larger than planets in our solar system. The Galilean moons of Jupiter - Io, Europa, Ganymede, and Callisto - are a remarkable ensemble, each with unique features and mysteries. Saturn's moon Titan, shrouded in a thick atmosphere and lakes of liquid hydrocarbons, has become a subject of particular interest for future exploration.

The Asteroid Belt: Remnants of Cosmic Evolution

Asteroids: Primordial Building Blocks

The asteroid belt is a fascinating region situated between the orbits of Mars and Jupiter, approximately 2.1 to 3.3 astronomical units (AU) away from the Sun. An astronomical unit is the average distance between the Earth and the Sun, about 93 million miles or 150 million kilometers. This vast region is home to a myriad of rocky and metallic bodies known as asteroids, which vary in size from tiny pebbles to large objects hundreds of miles in diameter.

Formation of the Asteroid Belt:

The formation of the asteroid belt is intertwined with the early history of our solar system. Around 4.6 billion years ago, a vast cloud of gas and dust, known as the solar nebula, began to collapse under the influence of gravity. As this nebula contracted, it started to spin and flatten into a rotating disk with the Sun at its center.

Within this protoplanetary disk, solid particles collided and accreted to form planetesimals - small rocky bodies ranging from a few meters to hundreds of kilometers in size. The most substantial planetesimals continued to grow by attracting nearby material through gravity, eventually becoming protoplanets.

However, the region between Mars and Jupiter proved to be a challenging environment for a large protoplanet to form. The gravitational influence of Jupiter, the largest planet in our solar system, prevented the accumulation of enough material to create another massive planet. Instead, the gravitational perturbations of Jupiter scattered or disrupted any growing protoplanets in this region.

Jupiter's Role in the Asteroid Belt:

Jupiter's immense gravity essentially acted as a cosmic barrier, halting the growth of any protoplanet that tried to emerge between itself and Mars. As a result, a significant number of planetesimals and smaller bodies were left in this region, forming what we now know as the asteroid belt.

While some of these asteroids are relatively small and irregularly shaped, others are large enough to have undergone differentiation, with heavier materials sinking to their centers, much like how planets develop their cores. However, the total mass of all the asteroids in the belt is only a tiny fraction of the mass of Earth's Moon, making the asteroid belt far less dense than commonly portrayed in science fiction.

Characteristics of Asteroids:

Asteroids in the belt exhibit a wide range of compositions, reflecting their diverse origins and histories. The two primary categories of asteroids in the belt are:

C-Type (Carbonaceous) Asteroids: These asteroids are the most common in the belt and are rich in carbon, as well as minerals such as clays and hydrated silicates. They are believed to be some of the most primitive and oldest objects in the solar system, dating back to its early stages.
S-Type (Silicaceous) Asteroids: These asteroids are rich in silicate minerals and are thought to be remnants of larger bodies that underwent some degree of heating and differentiation before being broken apart.

The asteroid belt is not a densely packed field of objects colliding with each other frequently. The average distance between asteroids is vast, and they are separated by millions of miles. Collisions do occur over long timescales, but they are relatively rare, with the probability of spacecraft encountering an asteroid in the belt being quite low.

Comets: Cosmic Wanderers

Comets: Icy Messengers from the Depths

Comets are celestial nomads, often dwelling in the farthest reaches of the solar system in regions called the Oort Cloud and the Kuiper Belt. When they journey close to the Sun, they develop magnificent tails of gas and dust, creating spectacular displays visible from Earth.

Our cosmic neighborhood is a diverse and dynamic playground of celestial wonders. From the fiery furnace of the Sun to the enigmatic icy realms of comets, each component contributes to the intricate dance of our solar system. As we continue to explore and understand these celestial bodies, we gain profound insights into the broader mysteries of the cosmos and our place within it.

A Detailed Look at Each Planet and Its Unique Features

In our cosmic neighborhood, each planet presents a captivating and distinctive world, shaped by its size, composition, atmosphere, and proximity to the Sun. Let's take a closer look at each planet in our solar system:

1. Mercury: The Swift Messenger

Mercury, the innermost planet, is a barren and rocky world, resembling Earth's Moon in many ways. Being the closest planet to the Sun, its surface experiences extreme temperature variations, with scorching daytime highs and freezing nighttime lows. Despite its proximity to the Sun, Mercury's lack of a significant atmosphere means that it has no insulation to retain heat.
Unique Features:

- Scarred Surface: Mercury's surface is heavily cratered, evidence of a long history of cosmic bombardment.
- Extreme Temperature Extremes: During the day, temperatures can soar up to 800 degrees Fahrenheit (430 degrees Celsius), while at night, they plummet to around -290 degrees Fahrenheit (-180 degrees Celsius).
- Lack of Atmosphere: Mercury has a thin exosphere composed mostly of atoms blasted off its surface by solar radiation and solar wind.

2. Venus: The Enigmatic Hellish World

Venus is often called Earth's "sister planet" due to its similar size and composition. However, that's where the resemblance ends. Venus has a thick atmosphere primarily composed of carbon dioxide, with clouds of sulfuric acid. This atmosphere traps heat, creating a runaway greenhouse effect that makes Venus the hottest planet in our solar system, even hotter than Mercury.

Unique Features:
- Runaway Greenhouse Effect: Venus has surface temperatures that can reach a blistering 900 degrees Fahrenheit (475 degrees Celsius), hotter than the surface of Mercury despite being farther from the Sun.
- Crushing Atmospheric Pressure: Venus's atmospheric pressure at the surface is about 92 times that of Earth, equivalent to the pressure found 900 meters deep in our oceans.
- Retrograde Rotation: Venus rotates on its axis in the opposite direction compared to most other planets, making its day longer than its year.

3. Earth: The Blue Marble

Our home planet, Earth, is teeming with life and abundant water, making it a unique oasis in the cosmos. With a protective atmosphere and a balanced climate, Earth nurtures

a remarkable diversity of ecosystems and supports a wide array of life forms.
Unique Features:

- Life-Supporting Atmosphere: Earth's atmosphere contains oxygen, a byproduct of photosynthesizing plants, allowing for the proliferation of complex life forms.
- Liquid Water: The presence of liquid water is essential for life as we know it, and Earth is the only planet in our solar system with vast oceans and liquid water on its surface.
- Diverse Ecosystems: Earth's varied environments, from lush rainforests to frozen polar regions, showcase the incredible biodiversity that thrives on our planet.

4. Mars: The Red Planet

Mars, often referred to as the "Red Planet," has intrigued scientists and science fiction enthusiasts alike due to its potential for past or present life. It is a cold, dry, and dusty world with a thin atmosphere, mainly composed of carbon dioxide.
Unique Features:

- Ancient Riverbeds: Mars shows evidence of ancient river valleys and dried-up lake beds, suggesting that liquid water once flowed on its surface.
- Seasonal Polar Ice Caps: Mars has polar ice caps made of water ice and dry ice (frozen carbon dioxide) that expand and contract with the changing seasons.
- Olympus Mons: Mars is home to Olympus Mons, the largest volcano in the solar system, reaching a height of over 13 miles (21 kilometers).

5. Jupiter: The King of Planets

Jupiter is the largest planet in our solar system, a gas giant composed mostly of hydrogen and helium. Its massive size and

powerful magnetic field make it a cosmic guardian, protecting the inner planets from potential comet and asteroid impacts.
Unique Features:

- Great Red Spot: Jupiter's most iconic feature is the Great Red Spot, a massive and long-lasting storm that has been raging for at least 350 years.
- Extensive System of Moons: Jupiter has a vast system of moons, with over 70 known moons, including the four Galilean moons: Io, Europa, Ganymede, and Callisto.
- Strong Magnetic Field: Jupiter has a powerful magnetic field, making it the most powerful magnetic planet in our solar system.

6. Saturn: The Ringed Wonder

Saturn, renowned for its spectacular ring system, is another gas giant in our solar system. Its mesmerizing rings are made up of countless small particles, ranging in size from micrometers to meters.
Unique Features:

- Magnificent Rings: Saturn's ring system is one of the most captivating sights in the night sky. The rings are made mostly of ice particles, with some containing rocky material.
- Titan: Saturn's largest moon, Titan, is a fascinating world with a thick atmosphere rich in nitrogen and methane. It has lakes and seas of liquid hydrocarbons on its surface.
- Cassini Mission: NASA's Cassini spacecraft provided detailed observations of Saturn and its moons, revealing their diverse features and mysteries.

7. Uranus: The Tilted Ice Giant

Uranus is an ice giant, consisting mainly of water, ammonia, and methane ices. It has a peculiar characteristic of rotating on its side, with its axis of rotation nearly parallel to its orbital plane.

Unique Features:

- Extreme Axial Tilt: Uranus's axial tilt is approximately 98 degrees, causing its poles to experience long periods of darkness and light during its 84-year-long orbit.
- Methane Atmosphere: Uranus's atmosphere contains methane, which absorbs red light, giving the planet its bluish-green hue.
- Unique Ring System: Uranus has a system of thin and dark rings that were discovered during a stellar occultation event.

8. Neptune: The Farthest Ice Giant

Neptune, the farthest planet from the Sun, is an ice giant similar in composition to Uranus. It has a striking blue color due to the presence of methane in its atmosphere.
Unique Features:

- The Great Dark Spot: Neptune exhibits atmospheric features similar to Jupiter's Great Red Spot, including a "Great Dark Spot" that was observed by the Voyager 2 spacecraft in 1989.
- Triton: Neptune's largest moon, Triton, is one of the coldest objects in our solar system and has a nitrogen-rich atmosphere.
- Dynamic Weather: Neptune experiences extreme weather patterns, with the fastest winds in the solar system, reaching speeds of up to 1,500 miles per hour (2,400 kilometers per hour).

Each planet in our solar system offers a distinctive and captivating perspective on the wonders of the cosmos. From the scorched plains of Mercury to the frigid moons of Neptune, each world has its story to tell and continues to inspire scientific inquiry and exploration. As we continue to study and explore these celestial bodies, we deepen our understanding of the solar system's evolution and the vast diversity of worlds that inhabit our cosmic neighborhood.

The Search for Exoplanets and Potential Habitable Worlds

For centuries, humanity has wondered whether other worlds beyond our solar system exist, and whether any of them could support life as we know it. This quest to find exoplanets - planets outside our solar system - and identify potential habitable worlds has become one of the most exciting and profound pursuits in modern astronomy.

The First Discoveries

The Pioneering Exoplanet: 51 Pegasi b (1995)

In 1995, Swiss astronomers Michel Mayor and Didier Queloz made history by discovering 51 Pegasi b, the first confirmed exoplanet orbiting a Sun-like star. This milestone find opened the floodgates for exoplanet exploration.

Kepler Space Telescope (2009 - 2018)

Launched in 2009, NASA's Kepler space telescope revolutionized the search for exoplanets by continuously monitoring a patch of the sky for the subtle dimming of stars caused by planets passing in front of them (transits). Kepler discovered thousands of exoplanets, providing essential data for characterizing these distant worlds.

Methods of Detection

Transit Method

The transit method involves observing the periodic dimming of a star's light as a planet passes in front of it. This change in brightness allows astronomers to infer the presence, size, and orbital characteristics of the exoplanet.

Radial Velocity (Doppler) Method

By detecting the slight "wobble" of a star caused by the gravitational tug of an orbiting planet, astronomers can determine the presence and mass of the exoplanet.

Direct Imaging

Using advanced telescopes and instruments, astronomers can capture images of exoplanets by blocking out the light from their host stars. This method allows for detailed observations of the exoplanet's properties.

Gravitational Microlensing

The gravitational microlensing technique relies on the bending of light by a massive object (like a star) to amplify the light of a more distant star. If a planet is orbiting the foreground star, its presence can be inferred from the distortion in the light curve.

Characteristics of Exoplanets

Hot Jupiters

Many early exoplanet discoveries were "Hot Jupiters" - gas giants like Jupiter but in close orbits around their stars. Their existence challenged traditional theories of planet formation and migration.

Super-Earths

Super-Earths are exoplanets with masses greater than Earth but less than Neptune. They come in various sizes and compositions and may represent a substantial fraction of exoplanets found to date.

The Quest for Habitable Worlds

The Habitable Zone

The habitable zone is the region around a star where conditions may be right for liquid water to exist on the surface of a rocky planet. This zone is not too hot, like Venus, nor too cold, like Mars.

Exoplanets in the Habitable Zone

Astronomers have discovered several exoplanets within the habitable zone of their stars. Examples include Proxima Centauri b, orbiting the nearest star to the Sun, and TRAPPIST-1 system, with seven Earth-sized planets.

Future Prospects

TESS Mission

NASA's Transiting Exoplanet Survey Satellite (TESS) is continuing the search for exoplanets using the transit method. TESS is expected to discover thousands of additional exoplanets during its mission.

James Webb Space Telescope (JWST)

Scheduled for launch in 2021, the JWST promises to revolutionize our understanding of exoplanets by studying their atmospheres and properties in unprecedented detail.

The search for exoplanets and potential habitable worlds continues to excite and inspire scientists and the public alike. Each discovery brings us closer to understanding our place in the cosmos and the possibility that life may exist beyond Earth. As technology advances and new missions are launched, the dream of finding an Earth-like exoplanet with signs of life becomes ever more within reach.

Chapter 3: Galactic Splendor

The Milky Way: Our Stellar Home

The Galactic Center

At the heart of our galaxy lies a mysterious region known as the Galactic Center. It is obscured from direct view by dust and gas, making observations in visible light challenging. However, infrared and radio observations have revealed the presence of a supermassive black hole called Sagittarius A* (Sgr A*), which likely harbors several million times the mass of our Sun.

Spiral Arms and Star Formation

The Milky Way is a barred spiral galaxy, characterized by its central bar-shaped structure and graceful spiral arms. These arms, such as the Perseus Arm and the Orion Arm, contain dense regions of gas and dust where new stars are continuously being born. Stellar nurseries within these arms give rise to young and massive stars, illuminating the galactic landscape.

Galactic Clusters and Nebulae

Open Star Clusters

Throughout the Milky Way, open star clusters grace the sky. These collections of young stars are held together by their mutual gravitational attraction and often shine brilliantly, showcasing the beauty of stellar birthplaces.

Globular Clusters

Globular clusters are densely packed groups of ancient stars that orbit around the galaxy's core. These ancient stellar relics provide a glimpse into the early history of the Milky Way and the universe itself.

Nebulae: Cosmic Artistry

Nebulae are vast clouds of gas and dust that come in various shapes and colors. Emission nebulae, such as the Orion Nebula, glow due to the ionization of gas by nearby hot stars.

Reflection nebulae, like the Pleiades, shimmer as they scatter light from nearby stars. Dark nebulae, such as the Horsehead Nebula, appear as eerie patches where dense clouds of dust block the light from stars beyond.

Galactic Dynamics and Halo

Galactic Rotation

The Milky Way spins gracefully, with stars in its outer regions orbiting the galactic center at different speeds. This differential rotation gives rise to the galaxy's spiral structure, as stars maintain their relative positions within the spiral arms.

Dark Matter Halo

While stars, gas, and dust are visible components of the Milky Way, a significant portion of its mass is attributed to dark matter. This mysterious and invisible substance exerts a gravitational influence, shaping the galaxy's overall structure and providing the "halo" around it.

Interstellar Phenomena

Supernova Remnants

Supernova explosions mark the spectacular deaths of massive stars, scattering heavy elements into space. The remnants of these explosions, called supernova remnants, enrich the interstellar medium with the building blocks of future stars and planetary systems.

Cosmic Rays

High-energy particles from various sources, including supernovae and active galactic nuclei, travel through space as cosmic rays. These particles play a crucial role in the dynamics and evolution of the interstellar medium.

The Galactic Halo and Beyond

Galactic Halo

Beyond the main disk of the Milky Way lies the galactic halo, a region of sparse stars and globular clusters that envelops the galaxy. The halo extends into the outskirts of the galaxy, where faint and ancient stars quietly populate the cosmic realm.

Intergalactic Space

In the vast expanse between galaxies lies the intergalactic space, a sparsely populated and mysterious realm. Intergalactic gas, dust, and dark matter weave a cosmic tapestry connecting galactic superstructures.

The Story of Our Galaxy

The Milky Way, our home galaxy, is an awe-inspiring tapestry of stars, gas, dust, and dark matter. Its story stretches back over billions of years, encompassing cosmic events that have shaped its structure, evolution, and the creation of everything we see around us today.

Galactic Formation

The origin of the Milky Way can be traced back to the early universe, approximately 13.8 billion years ago. As the universe expanded and cooled, tiny fluctuations in density led to the formation of structures, including galaxies. Gravity drew matter together, and over time, clumps of gas and dust coalesced to form the first stars and galaxies.

The Milky Way likely formed from the merger and accretion of smaller protogalactic structures. As gas and dust collapsed under their mutual gravitational attraction, the first generation of stars ignited, beginning a new era of stellar evolution.

Galactic Structure

The Milky Way is a barred spiral galaxy, a shape that arises from the combined rotation of stars and the presence of a central bar-like structure. It consists of several key components:

Galactic Core: At the heart of the Milky Way lies a dense and mysterious region known as the Galactic Core. Hidden behind thick clouds of dust, this region is home to a supermassive black hole called Sagittarius A* (Sgr A*). The gravitational influence of this black hole shapes the orbits of stars and other celestial objects in its vicinity.

Galactic Disk: Surrounding the core is the Galactic Disk, a flattened region where most of the Milky Way's stars, gas, and dust reside. The beautiful spiral arms, such as the Perseus Arm and the Orion Arm, are part of the disk.

Halo: Extending beyond the disk is the Galactic Halo, a faint and sparsely populated region containing globular clusters and older stars. The halo is also believed to contain a significant amount of dark matter, an invisible substance that exerts gravitational influence on the galaxy.

Stellar Populations

The Milky Way hosts a diverse array of stars, each with its own age, composition, and evolutionary history. Stellar populations can be broadly classified into three main groups:

Population I Stars: These are relatively young stars that are rich in heavy elements. They are found primarily in the Galactic Disk and include stars like our Sun. Population I stars are associated with ongoing star formation and the presence of dust and gas.

Population II Stars: These are older stars that formed from gas with fewer heavy elements. Population II stars are found in the Galactic Halo and globular clusters, and they provide valuable insights into the early stages of the galaxy's evolution.

Population III Stars: The earliest generation of stars that formed in the universe, Population III stars, is thought to have been massive and short-lived. While no Population III stars have been directly observed, their existence is inferred from the abundance of heavy elements in later generations of stars.

Galactic Evolution

Over billions of years, the Milky Way has undergone significant changes driven by processes such as star formation, stellar evolution, and interactions with other galaxies. These events have shaped the galaxy's structure and enriched it with heavy elements, which are essential for the formation of planets, and ultimately, life.

Cosmic Recycling

Throughout its history, the Milky Way has experienced a continuous cycle of stellar birth, evolution, and death. Stars form from the gas and dust of molecular clouds, shine brightly throughout their main-sequence lifetime, and eventually exhaust their nuclear fuel. Massive stars end their lives in brilliant supernova explosions, scattering heavy elements into space. These enriched materials mix with interstellar gas, providing the building blocks for the formation of new stars and planetary systems.

The Milky Way's Place in the Universe

The Milky Way is not an isolated island in the cosmos. It belongs to a larger cosmic web, a vast network of galaxies gravitationally bound together. These galaxies come together to form galaxy clusters and superclusters, creating the grand structure of the universe. Understanding the story of our galaxy, the Milky Way, is a testament to human curiosity and our desire to explore the cosmos.

The Milky Way, with its myriad stars, clusters, and nebulae, tells a cosmic story that spans billions of years. As we unravel its mysteries, we gain deeper insights into the universe's evolution and our place in the grand tapestry of the cosmos. Our galaxy's splendor continues to captivate and inspire, inviting us to embark on a journey of cosmic exploration and discovery.

As we continue to peer deeper into the night sky, new discoveries await, enriching our knowledge of our place in the universe and the interconnectedness of all cosmic phenomena. The journey of unraveling the story of our galaxy has only just begun, and with each new revelation, we gain a deeper appreciation for the grandeur and splendor of the Milky Way.

Exploring the Structure and Composition of Galaxies

Galaxies are vast cosmic systems, each containing billions to trillions of stars, gas, dust, and dark matter, bound together by gravity. They come in various shapes and sizes, and exploring their structure and composition has been a central focus of astronomy for centuries. As our observational techniques and technology have advanced, we have gained profound insights into the diverse and intricate nature of galaxies.

Galaxy Morphology

Astronomers classify galaxies based on their visual appearance, known as galaxy morphology. The most common types of galaxies are:

Spiral Galaxies: These galaxies have prominent spiral arms that emanate from a central bulge. The Milky Way is a prime example of a barred spiral galaxy, where a bar-shaped structure extends from the core.

Elliptical Galaxies: Elliptical galaxies lack the distinct spiral arms and appear as elliptical or nearly spherical shapes. They range from nearly spherical (E0) to highly elongated (E7) in appearance.

Lenticular Galaxies: Lenticular galaxies have a disk-like structure similar to spiral galaxies but lack the prominent spiral arms. They are often described as intermediate between spiral and elliptical galaxies.

Irregular Galaxies: Irregular galaxies lack a well-defined shape and structure, often displaying chaotic and irregular patterns. They may result from gravitational interactions or mergers between galaxies.

Galactic Components

Galaxies are composed of various components, each contributing to their overall structure and behavior:

Stellar Population: The stars in a galaxy form different populations based on their age and metallicity (abundance of elements heavier than hydrogen and helium). Stellar populations provide insights into the galaxy's evolutionary history.

Interstellar Medium (ISM): The ISM comprises gas (mostly hydrogen) and dust between stars. It serves as the raw material for star formation and plays a crucial role in shaping a galaxy's dynamics.

Dark Matter: Dark matter is an invisible and mysterious form of matter that does not emit, absorb, or reflect light. It exerts gravitational influence, providing additional mass that shapes the galaxy's overall structure, including the dark matter halo.

Galactic Center and Nucleus

The center of many galaxies contains a dense region called the galactic center or nucleus. This region often hosts a supermassive black hole, millions to billions of times more massive than our Sun. The gravitational influence of the black hole can profoundly affect the surrounding stars and gas, leading to intense radiation and the formation of structures like accretion disks.

Galactic Evolution

Galaxies are not static entities; they undergo a continuous process of evolution shaped by various factors, including mergers with other galaxies, star formation rates, and interactions with the intergalactic medium.

Galactic Collisions and Mergers: When galaxies interact and collide, their structures can be dramatically altered. Mergers between galaxies lead to the formation of larger galaxies, such as ellipticals, and can trigger bursts of star formation.

Star Formation and Stellar Evolution: As stars form, evolve, and eventually die, they release energy and heavy elements back into the interstellar medium. These processes contribute to the galaxy's chemical enrichment and the formation of new stars.

Galactic Winds and Feedback: Energetic processes, such as supernovae and supermassive black hole activity, can drive powerful galactic winds. These winds expel gas from galaxies, affecting their future star formation and evolution.

Active Galactic Nuclei (AGN)

Some galactic centers host active galactic nuclei (AGN), regions where the supermassive black hole actively accretes matter. AGN emits prodigious amounts of energy across the electromagnetic spectrum and can profoundly influence the surrounding galaxy's evolution.

Galaxy Surveys and Observations

Advancements in observational astronomy have allowed us to conduct extensive surveys of galaxies across the universe. Surveys like the Sloan Digital Sky Survey (SDSS) and the Hubble Space Telescope's deep fields have provided vast catalogs of galaxies, offering crucial data for understanding their distribution, properties, and evolution.

The Search for Extragalactic Planets

While most of our focus has been on planets within our own solar system, astronomers are also actively searching for exoplanets beyond the Milky Way. These hypothetical "extragalactic planets" could exist in orbit around stars in other galaxies, further broadening our understanding of planetary systems.

Studying the structure and composition of galaxies continues to be a captivating field of research, leading to groundbreaking discoveries and deepening our understanding of the cosmos. As technology advances and we explore more distant regions of the universe, the story of galaxies, from their birth to their eventual fate, unfolds before our eyes, inviting us to delve deeper into the cosmic tapestry.

Black Holes, Quasars, and Other Cosmic Phenomena

Black Holes: Gravity's Abyss

Black holes are enigmatic cosmic objects with gravitational fields so intense that nothing, not even light, can escape their grasp. They form from the remnants of massive stars that have exhausted their nuclear fuel and undergo gravitational collapse.

Event Horizon: The boundary of a black hole, known as the event horizon, marks the point of no return. Once anything crosses this boundary, it is inexorably pulled into the black hole's core.

Singularity: At the center of a black hole lies the singularity, where gravitational forces become infinite and space-time curves infinitely. Our current understanding of physics breaks down at this point.

Black Hole Types: Black holes come in different sizes:

- Stellar Black Holes: Formed from the remnants of massive stars, typically a few times the mass of our Sun.
- Intermediate Black Holes: Rarer and more massive than stellar black holes, yet smaller than supermassive black holes.
- Supermassive Black Holes: Found at the centers of most galaxies, with masses ranging from millions to billions of times that of the Sun.

Quasars: Cosmic Beacons

Quasars (quasi-stellar radio sources) are exceptionally bright and distant objects powered by supermassive black holes actively accreting matter at their centers. These cosmic beacons are some of the most luminous and energetic phenomena in the universe.

Active Galactic Nuclei (AGN): Quasars are a type of AGN, where intense radiation and jets of particles are produced as matter falls into the supermassive black hole.

Light Travel: Some quasars are so distant that their light has taken billions of years to reach us, providing astronomers with glimpses into the early universe.

Gamma-Ray Bursts (GRBs): Cosmic Explosions

Gamma-ray bursts are brief but incredibly powerful bursts of gamma-ray radiation. They are among the most energetic events in the universe, originating from various sources, including the collapse of massive stars or the merger of neutron stars or black holes.

Short and Long GRBs: GRBs are classified into short-duration bursts (a few milliseconds to seconds) and long-duration bursts (lasting several seconds to minutes), indicating different origins.

Afterglows: After the initial gamma-ray burst, an afterglow in other wavelengths (X-ray, optical, radio) may persist for hours, days, or even longer, providing valuable data for understanding the phenomenon.

Neutron Stars and Pulsars: Cosmic Spinners

Neutron stars are the remnants of massive stars that have undergone supernova explosions. They are incredibly dense, with the mass of a few suns packed into a sphere the size of a city.

Pulsars: Neutron stars with powerful magnetic fields emit beams of radiation from their magnetic poles. As they rotate, these beams sweep across space, resulting in periodic flashes of radiation observable from Earth, called pulsars.

Millisecond Pulsars: Some pulsars spin at incredibly fast rates, with periods of only a few milliseconds. These millisecond pulsars are thought to be the result of the transfer of mass from a companion star.

Supernovae: Stellar Explosions

Supernovae are the cataclysmic explosions of massive stars. They release immense amounts of energy, outshining entire galaxies for brief periods.

Type I and Type II Supernovae: Supernovae are broadly categorized into Type I, resulting from the detonation of a white dwarf in a binary system, and Type II, the explosive death of massive stars.

Heavy Element Production: Supernovae are crucial for the synthesis and dispersal of heavy elements (like iron, gold, and uranium) into space, enriching the interstellar medium.

Cosmic Filaments and Large-Scale Structure

Cosmic filaments are vast, thread-like structures composed of galaxies, gas, and dark matter. They span hundreds of millions of light-years, connecting galaxy clusters and forming the large-scale structure of the universe.

Cosmic Web: The cosmic web is a network of filaments and voids that outline the distribution of galaxies and matter in the universe, forming an intricate and interconnected structure.

Dark Matter's Role: Dark matter provides the gravitational scaffolding for the formation and evolution of cosmic filaments and the overall large-scale structure of the universe.

Cosmic phenomena like black holes, quasars, gamma-ray bursts, neutron stars, supernovae, and cosmic filaments are captivating and mysterious. They offer unique windows into the workings of the universe, pushing the boundaries of our understanding of fundamental physics and the cosmos. As we continue to explore and study these cosmic wonders, we gain deeper insights into the complexity, beauty, and diversity of the universe we call home.

Chapter 4: The Great Beyond: Intergalactic Marvels

The Cosmic Web: A Tapestry of Galaxies

The universe is an expansive canvas, and at its grandest scale, galaxies are not isolated islands but part of a vast, interconnected structure known as the cosmic web. This web-like pattern comprises galaxies, galaxy clusters, filaments, and immense voids, woven together by the gravitational dance of dark matter.

Cosmic Voids: Voids are vast regions of the universe with relatively sparse matter. These seemingly empty spaces between cosmic filaments are vital for understanding the large-scale structure of the cosmos.

Galaxy Clusters: Galaxy clusters are the largest gravitationally bound structures in the universe, containing hundreds to thousands of galaxies. These clusters are veritable universes in their own right, influencing their member galaxies through gravitational interactions.

Superclusters: Superclusters are collections of multiple galaxy clusters, forming some of the most massive and intricate structures in the cosmic web. They serve as signposts of the universe's evolution and are fundamental in tracing the distribution of matter on a vast scale.

Intergalactic Gas and Galaxy Formation

The vast regions between galaxies are not empty but permeated by intergalactic gas—primarily composed of hydrogen and helium. This gas plays a pivotal role in galaxy formation and evolution.

Cosmic Filaments as Gas Highways: Intergalactic gas flows along cosmic filaments, feeding galaxies and facilitating the formation of new stars.

Feedback and Quenching: Intense radiation from active galactic nuclei (AGN) and supernovae explosions can heat and expel gas from galaxies, regulating star formation and influencing their growth.

Gravitational Lensing: Cosmic Mirages

Massive objects, such as galaxy clusters, can bend the path of light from distant objects behind them. This phenomenon, known as gravitational lensing, acts as a cosmic magnifying glass, allowing astronomers to observe distant galaxies with unprecedented clarity.
Strong Lensing: In strong lensing, the light from a distant galaxy is significantly distorted, creating arcs, rings, and multiple images around the foreground lensing object.
Weak Lensing: Weak lensing results in subtle distortions of background galaxies, providing valuable information about the distribution of dark matter in galaxy clusters.

Intergalactic Magnetic Fields

Beyond the galaxies and cosmic structures, magnetic fields pervade the vast cosmic voids. These intergalactic magnetic fields, while extremely weak, may play a crucial role in galaxy formation, influencing cosmic structure, and could offer insights into the early universe.

Cosmic Microwave Background (CMB)

The Cosmic Microwave Background is the faint glow of radiation left over from the Big Bang. It serves as a snapshot of the universe when it was just 380,000 years old.

Temperature Anisotropies: Tiny fluctuations in the CMB's temperature reveal the seeds of cosmic structure that eventually gave rise to galaxies and galaxy clusters.
Cosmological Parameters: Detailed analysis of the CMB has provided crucial data for determining the universe's age, composition, and expansion rate.

Intergalactic Travel and SETI

While intergalactic travel remains firmly in the realm of science fiction, the Search for Extraterrestrial Intelligence (SETI) searches for signals from advanced civilizations in other galaxies.

The Fermi Paradox: Considering the vast number of galaxies and potential habitable planets, the Fermi Paradox raises questions about the apparent absence of extraterrestrial civilizations.
Interstellar Communication: Scientists explore various methods of interstellar communication, including beamed signals and possible signatures of advanced technologies.

Multiverse Hypothesis

Beyond our observable universe, theoretical physicists speculate about the existence of a multiverse—a vast ensemble of universes with different physical laws and properties.
Inflationary Cosmology: The concept of inflation suggests that our universe rapidly expanded from an infinitesimal point, leading to the existence of multiple "bubble" universes.
String Theory and the Landscape: String theory proposes multiple possible solutions, leading to the idea of a landscape of universes with diverse physical constants.

The Great Beyond beckons with its intergalactic marvels, offering glimpses into the intricate web of cosmic structure, magnetic mysteries, and the cosmic microwave echoes of our universe's birth. While intergalactic travel may remain elusive, our pursuit of knowledge, exploration, and the quest for life beyond our home galaxy continue to drive us forward, unlocking the secrets of the vast and captivating cosmos.

The Vastness of the Universe Beyond Our Galaxy

The universe is a boundless expanse of cosmic wonders that extends far beyond the borders of our own Milky Way galaxy. Exploring the vastness of the cosmos reveals a staggering array of galaxies, galaxy clusters, and intergalactic phenomena that highlight the sheer scale and grandeur of the universe.

The Local Group: A Neighborhood of Galaxies

The Milky Way is not alone in the cosmic wilderness. It is part of a small cluster of galaxies known as the Local Group. This grouping includes approximately 54 galaxies, dominated by the Milky Way and Andromeda galaxies, along with numerous dwarf galaxies.

Andromeda (M31): The Andromeda Galaxy is the closest spiral galaxy to the Milky Way and offers a glimpse of our own galaxy's future fate, as the two are on a collision course that will eventually merge them into a larger elliptical galaxy.

Triangulum (M33): Also known as the Triangulum Galaxy, M33 is another prominent member of the Local Group, though it is smaller and less massive than the Milky Way and Andromeda.

The Expanse of Cosmic Distances

The scale of intergalactic distances is mind-boggling. Even within the Local Group, the average separation between galaxies is millions of light-years. Traveling from one galaxy to another would take lifetimes, even at the speed of light.

Cosmic Units of Distance: To cope with these vast distances, astronomers use units such as light-years, where one light-year is the distance light travels in one year (about 9.46 trillion kilometers or 5.88 trillion miles).

Observable Universe: The observable universe is the portion of the universe we can currently detect, estimated to be about 93 billion light-years in diameter.

Cosmic Structure: Galaxy Clusters and Superclusters

At even larger scales, galaxies clump together to form galaxy clusters and superclusters, creating immense cosmic structures.

Galaxy Clusters: Galaxy clusters are some of the largest structures in the universe, containing thousands of galaxies bound together by gravity.

Superclusters: Superclusters are vast assemblages of galaxy clusters, spanning hundreds of millions of light-years. The Great Attractor is one such mysterious supercluster that influences the motion of galaxies in its vicinity.

The Cosmic Web: A Latticework of Galaxies

The large-scale structure of the universe resembles a cosmic web, with galaxy filaments crisscrossing immense voids. This complex network of galaxies is shaped by the gravitational pull of dark matter.

Dark Matter Dominance: Dark matter, an invisible and mysterious form of matter, constitutes about 27% of the universe's mass and significantly influences the formation of the cosmic web.

Voids and Cosmic Filaments: Voids, vast regions with relatively few galaxies, separate the intricate filaments, where galaxies are more densely clustered.

Intergalactic Gas and Cosmic Evolution

The vast cosmic voids are not empty but filled with intergalactic gas. This gas plays a crucial role in the cosmic evolution, feeding galaxy formation and influencing the growth of cosmic structures.

Intergalactic Medium (IGM): The IGM is composed mostly of hydrogen and helium left over from the Big Bang, as well as heavier elements produced by stars.
Cosmic Reionization: The early universe was dominated by neutral gas. As the first stars and galaxies formed, they emitted intense radiation that ionized the gas, leading to the "cosmic reionization" epoch.

Cosmic Expansion: The Fate of the Universe

The universe is not static but continuously expanding. The fate of this expansion depends on the balance between the universe's mass and the energy associated with dark energy.

Cosmic Expansion Rate: The expansion rate of the universe, quantified by the Hubble constant, governs the rate at which galaxies move apart from one another.
Future of the Universe: Depending on the density of matter and dark energy, the universe could experience a "Big Freeze," "Big Rip," or "Big Crunch" scenario, each with vastly different implications for the cosmos' fate.

The vastness of the universe beyond our galaxy invites us to ponder the mysteries of cosmic structure, the influence of dark matter and dark energy, and the ultimate fate of the cosmos. As we continue to explore and unravel the secrets of the Great Beyond, we gain a deeper appreciation for the interconnectedness of all cosmic phenomena and the awe-inspiring beauty of the cosmos that stretches beyond the horizon of human imagination.

Clusters, Superclusters, and Cosmic Voids: Cosmic Tapestry

Galaxy Clusters: Cosmic Cities

Galaxy clusters are among the most massive and awe-inspiring structures in the universe. These cosmic metropolises contain hundreds to thousands of galaxies bound together by gravity.

Gravitational Dance: Within a galaxy cluster, galaxies are engaged in a gravitational dance, orbiting their common center of mass. The cluster's immense gravity shapes the motion and distribution of galaxies within it.
Intracluster Medium: The space between galaxies in a cluster is filled with hot, diffuse gas known as the intracluster medium. This gas, primarily composed of hydrogen and helium, emits X-rays and serves as the repository of baryonic matter.

Superclusters: Vast Cosmic Assemblages

Superclusters are the next level of cosmic organization, encompassing multiple galaxy clusters and spanning hundreds of millions of light-years.

Connecting Galaxies: Superclusters act as bridges between galaxy clusters, linking them together through vast cosmic filaments. These filaments form the backbone of the cosmic web.
Great Attractor: The Great Attractor is a mysterious supercluster that influences the motion of galaxies in its vicinity, including our Local Group of galaxies.

Cosmic Voids: Vast Expanses

Cosmic voids are immense regions of the universe with relatively sparse matter. They are the antithesis of galaxy clusters and superclusters.

Vast and Empty: Voids can be tens of millions of light-years in diameter and contain significantly fewer galaxies than other regions of the universe.

Expansion of the Universe: While galaxies and clusters are bound together by gravity, the cosmic voids are expanding along with the overall expansion of the universe.

The Large-Scale Structure: A Cosmic Web

The universe's large-scale structure resembles a cosmic web—a vast latticework of galaxy clusters, superclusters, and cosmic voids. This intricate web is shaped by the gravitational pull of dark matter, the invisible and mysterious substance that makes up a significant portion of the universe's mass.

Cosmic Filaments: The filaments that compose the cosmic web are vast threads of galaxies that stretch across immense distances. These filaments are the highways along which galaxies travel.

Voids in the Web: The spaces between filaments are the cosmic voids, where galaxies are scarce, and matter is relatively diffuse.

Cosmic Evolution: Past and Future

The distribution of galaxy clusters, superclusters, and voids is not static but evolves over cosmic timescales.

Formation and Growth: Galaxy clusters and superclusters form through the gravitational attraction of matter, while cosmic voids expand as the universe itself expands.

Cosmic Web and Dark Matter: Dark matter is instrumental in shaping the cosmic web, providing the gravitational scaffolding around which galaxies and galaxy clusters assemble.

Observing the Cosmic Web

Studying the large-scale structure of the universe is an essential aspect of modern cosmology.

Galaxy Surveys: Astronomers conduct extensive galaxy surveys to map the distribution and properties of galaxies, revealing the intricate structure of the cosmic web.

Cosmic Microwave Background: The faint glow of the cosmic microwave background (CMB) radiation carries imprints of the early universe's density fluctuations, providing crucial data for understanding the large-scale structure.

The clustering of galaxies, the vast expanse of superclusters, and the vast cosmic voids together form the tapestry of the cosmic web. Unraveling this cosmic tapestry offers profound insights into the evolution of the universe, the nature of dark matter and dark energy, and our place in the vast and interconnected cosmos. As we continue to explore the mysteries of clusters, superclusters, and cosmic voids, we gain a deeper appreciation for the beauty and complexity of the cosmic web that surrounds us.

Unraveling the Mysteries of Dark Matter and Dark Energy

The cosmos is a realm of profound mysteries, and two of its most enigmatic components are dark matter and dark energy. Although invisible and elusive, these entities exert significant influences on the universe's structure, expansion, and evolution. As scientists strive to understand these cosmic enigmas, they find themselves immersed in a quest that could revolutionize our understanding of the universe itself.

Dark Matter: The Invisible Gravitational Architect

Missing Mass Puzzle: Observations of galaxies and galaxy clusters revealed discrepancies between their visible mass (stars, gas, and dust) and the total mass inferred from gravitational effects. Dark matter was proposed as a solution to this "missing mass" puzzle.

Dark Matter Candidates: While dark matter remains undetected, various hypothetical particles, such as Weakly Interacting Massive Particles (WIMPs) and Axions, have been proposed as potential dark matter candidates.

Cosmic Evidence: Dark matter's gravitational influence can be seen in the rotational curves of galaxies, the gravitational lensing of distant objects, and the large-scale distribution of matter in the cosmic web.

Structure Formation: Dark matter played a crucial role in the formation of cosmic structures, acting as gravitational scaffolding for the clumping of matter and the birth of galaxies and galaxy clusters.

Dark Energy: The Accelerating Expander

Cosmic Acceleration: Observations of distant supernovae revealed that the universe's expansion is accelerating, contrary to expectations based on gravity's attractive nature. Dark energy was proposed as the source of this acceleration.

Cosmological Constant: Dark energy is often associated with the cosmological constant, a term introduced by Albert Einstein in his equations of general relativity.

Repulsive Force: Dark energy exerts a repulsive force that counteracts gravity's attractive pull on cosmic scales, driving the universe's accelerated expansion.

Cosmic Fate: The ultimate fate of the universe depends on dark energy's properties. If dark energy remains constant or weakly varies with time, the universe could experience an eternal expansion known as the "Big Freeze."

Unraveling the Mysteries

Cosmological Probes: Astronomers use various cosmological probes, such as the cosmic microwave background, galaxy surveys, and gravitational lensing, to study the properties and effects of dark matter and dark energy.

Large-Scale Simulations: High-performance computing allows researchers to simulate the behavior of dark matter and ordinary matter on cosmic scales, helping to compare theoretical predictions with observations.

Laboratory Experiments: Scientists conduct experiments deep underground or in particle accelerators to search for dark matter particles or evidence of their interactions.

The Dark Energy Survey (DES): DES is a large astronomical survey that aims to probe dark energy's nature by studying supernovae, galaxy clustering, and weak gravitational lensing.

The Cosmic Frontier

Unraveling the mysteries of dark matter and dark energy remains one of the greatest scientific endeavors of our time. As we journey into the cosmic frontier, we are compelled to question the very fabric of the universe. By understanding these enigmatic entities, we aspire to unlock the secrets of cosmic evolution, the fate of the universe, and our place in the vast cosmic tapestry. The quest for dark matter and dark energy inspires us to push the boundaries of knowledge, bringing us closer to comprehending the fundamental nature of the cosmos itself.

Chapter 5: Alien Worlds: A Search for Life

The search for life beyond our planet is a fascinating and profound scientific quest that captivates the human imagination. As we explore the cosmos, we seek to answer one of humanity's most profound questions: Are we alone in the universe? This chapter delves into the diverse methods and discoveries in the search for alien worlds and the potential for extraterrestrial life.

The Habitable Zone: Goldilocks for Life

Defining the Habitable Zone: The habitable zone, also known as the "Goldilocks zone," refers to the region around a star where conditions are just right for liquid water to exist on the surface of a planet. This zone is crucial for the potential development of life as we know it.

Exoplanets in the Habitable Zone: Astronomers have identified numerous exoplanets located in their star's habitable zones. These exoplanets come in a variety of sizes and orbits, offering tantalizing targets for further study.

Exoplanets: Alien Worlds Revealed

The Exoplanet Revolution: The discovery of exoplanets has revolutionized our understanding of the cosmos. Thanks to space missions like Kepler, TESS, and ground-based observatories, thousands of exoplanets have been detected, each offering unique characteristics and potential for life.

Exoplanet Detection Methods: Astronomers use various techniques, such as the transit method (observing a planet passing in front of its star), radial velocity method (detecting the star's wobble due to the planet's gravitational pull), and direct imaging, to identify exoplanets.

Exoplanet Atmospheres: Clues to Habitability

Spectroscopy: Analyzing the light passing through an exoplanet's atmosphere enables scientists to study its chemical composition, revealing essential molecules like water vapor, methane, and carbon dioxide.

Biosignatures: Certain atmospheric compositions, such as an abundance of oxygen and methane together, could be indicative of biological activity on an exoplanet.

Extreme Life on Earth: A Window to Alien Life?

Extreme Environments: Earth's extreme environments, such as deep-sea hydrothermal vents, acidic hot springs, and frozen polar regions, provide analogs for potential extraterrestrial habitats.

Extremophiles: Microorganisms known as extremophiles have adapted to thrive in extreme conditions on Earth, raising the possibility that life could exist in similarly harsh environments on other worlds.

SETI: Listening for Extraterrestrial Signals

The Search for Extraterrestrial Intelligence (SETI): Scientists use radio telescopes and other technologies to scan the skies for potential signals from advanced extraterrestrial civilizations.

The Wow! Signal: One of the most famous SETI incidents is the Wow! signal, a mysterious radio signal detected in 1977, which has yet to be conclusively explained.

Mars: The Search for Martian Life

Mars Exploration: Multiple missions to Mars, including rovers and landers, continue to search for evidence of past or present life on the Red Planet. Recent discoveries of liquid water beneath the Martian surface have rekindled interest in the possibility of Martian life.

Mars Sample Return: Upcoming missions aim to collect Martian soil and rock samples and return them to Earth for in-depth analysis, which could provide critical insights into the planet's habitability.

Beyond Mars: Ocean Worlds and Exomoons

Ocean Worlds: Moons of giant planets, such as Europa (Jupiter) and Enceladus (Saturn), are intriguing targets in the search for life. Beneath their icy surfaces, subsurface oceans could harbor the conditions for life.

Exomoons: Moons orbiting exoplanets could also provide habitable environments, expanding the possibilities for life beyond just planets.

The Drake Equation: Estimating the Prevalence of Life

The Drake Equation is a mathematical formula used to estimate the number of potential extraterrestrial civilizations in our galaxy based on factors such as the rate of star formation, the number of habitable planets, and the likelihood of life emerging.

The Future of the Search

Advancements in technology and future space missions hold the promise of further unlocking the secrets of alien worlds and the potential for life beyond Earth.

James Webb Space Telescope (JWST): The JWST, set to launch, will be capable of studying exoplanet atmospheres and searching for biosignatures.

The Next Generation of Exoplanet Missions: Planned missions like PLATO and ARIEL aim to broaden our understanding of exoplanets and their potential for habitability.

As we explore the cosmos, the search for alien worlds and signs of extraterrestrial life is a compelling journey that sparks curiosity and wonder. From exoplanets in distant star systems to the icy moons of our own solar system, each discovery brings us closer to unraveling the mystery of life beyond Earth. The search for alien life not only expands our knowledge of the universe but also challenges our perception of our place within it, inspiring us to contemplate the potential diversity and interconnectedness of life in the cosmos.

The Concept of Extraterrestrial Life and Its Possibilities

The concept of extraterrestrial life has long captured the human imagination, inspiring countless works of science fiction and sparking scientific curiosity about the potential existence of life beyond Earth. As we venture into the vast expanse of the universe, the possibilities of extraterrestrial life ignite our sense of wonder and open up a realm of questions about our place in the cosmos.

The Diversity of Life on Earth

Life's Resilience: Earth's extreme environments, from scorching deserts to freezing polar regions, showcase the resilience and adaptability of life. This diversity underscores the potential for life to thrive in diverse and even harsh conditions on other worlds.

Extremophiles: Microorganisms known as extremophiles have been found in extreme environments on Earth, such as hot springs, acidic lakes, and deep-sea hydrothermal vents, suggesting that life might persist in similarly challenging environments elsewhere.

Habitability and the Habitable Zone

The Habitable Zone Revisited: The concept of the habitable zone is not limited to planets orbiting stars. Moons orbiting gas giants and even rogue planets adrift in space could host conditions conducive to life.

Ocean Worlds: Moons like Europa (Jupiter) and Enceladus (Saturn), with subsurface oceans beneath their icy crusts, are promising targets for the search for life beyond Earth.

Mars: The Historical Focus of the Search

Historic Interest: Mars has been a focal point in the search for extraterrestrial life due to its proximity and potential for ancient habitability.

Water on Mars: Evidence of past water flows and recent discoveries of liquid water beneath the surface have fueled the idea that Mars might have been habitable in the past or could still harbor life today.

Exoplanets: A Universe of Possibilities

Exoplanet Discoveries: The detection of thousands of exoplanets, many in their star's habitable zone, highlights the diversity of planetary systems in our galaxy and the potential for Earth-like worlds.

Atmospheric Clues: Studying the atmospheres of exoplanets provides crucial information about their potential habitability and the presence of biosignatures.

The Drake Equation Revisited

The Drake Equation serves as a tool for contemplating the probability of extraterrestrial civilizations, considering factors such as star formation rates, the fraction of stars with planets, and the likelihood of life emerging.

Fermi Paradox: The Search for Intelligent Life

The Fermi Paradox poses the question of why, despite the vast number of potentially habitable planets, we have not yet observed any clear signs of intelligent extraterrestrial civilizations.

The Challenges of Communication

The Great Silence: The challenges of communication across interstellar distances and the vastness of space could be significant factors in the apparent lack of contact with extraterrestrial civilizations.

Interstellar Travel: The vast distances between stars present significant obstacles to interstellar travel, raising questions about the feasibility of physically reaching other habitable worlds.

The Nature of Alien Life

Biological Diversity: Life beyond Earth may differ significantly from terrestrial life, leading to speculation about possible non-carbon-based life forms or organisms based on different biochemistries.

Technological Civilizations: The search for extraterrestrial life extends beyond biology to the possibility of advanced civilizations capable of transmitting signals or visiting other star systems.

The Societal Impact of Discovery

The discovery of extraterrestrial life, even microbial, could have profound societal implications, reshaping our understanding of life, religion, and our place in the universe.

The concept of extraterrestrial life fuels our sense of curiosity and ignites our sense of exploration as we search the cosmos for signs of life beyond our home planet. While many questions remain unanswered, the possibilities of finding alien life, whether simple or complex, inspire us to explore, challenge our perceptions, and invite us to contemplate the vastness of the cosmos and our place within it. As the search continues, each discovery, be it the detection of a distant exoplanet or the exploration of an icy moon, brings us one step closer to unraveling the mystery of whether life exists beyond Earth.

Examining the Conditions for Life on Other Planets

The search for life beyond Earth is a multidisciplinary endeavor that involves astrophysics, astrobiology, and planetary science. As we explore exoplanets and other celestial bodies, we carefully examine various conditions that could support or hinder the development of life as we know it. Understanding these factors is crucial in the search for habitable environments and potential extraterrestrial life.

1. Presence of Liquid Water

The Universal Solvent: Water is essential for life as we know it. The presence of liquid water on a planet's surface or subsurface greatly enhances its potential for supporting life.
The Goldilocks Zone: Planets in their star's habitable zone, where temperatures are just right for liquid water to exist, are prime targets in the search for habitable exoplanets.

2. Atmosphere and Climate

Greenhouse Effect: An atmosphere capable of retaining heat through a greenhouse effect is crucial for maintaining stable temperatures conducive to life.
Protection from Radiation: A planet's atmosphere should shield its surface from harmful cosmic and stellar radiation that could otherwise be detrimental to life.

3. Presence of Essential Chemical Elements

Carbon, Oxygen, and Nitrogen: Elements like carbon, oxygen, and nitrogen are essential building blocks for organic molecules, which are the basis of life on Earth.
Trace Elements: The presence of trace elements, such as phosphorus and sulfur, is also vital for biological processes.

4. Energy Sources

Stellar Radiation: For life as we know it, energy from the host star (like solar energy from the Sun) is essential for driving biological processes through photosynthesis or other energy-conversion mechanisms.
Chemical Energy: On certain celestial bodies like hydrothermal vents or subsurface oceans, chemical energy from interactions between rocks and water can support life in the absence of sunlight.

5. Stability and Longevity

Stability of Conditions: Habitability depends on the stability of environmental conditions over geological timescales, allowing for the potential evolution and sustainability of life.
Longevity of Habitability: Planets or moons with long periods of habitability provide more opportunities for life to emerge and evolve.

6. Protection from Cosmic Hazards

Planetary Magnetic Fields: A strong magnetic field can shield a planet's atmosphere from erosion by the solar wind and cosmic rays.
Impact Protection: A planet with a stable orbit and protection from frequent and catastrophic asteroid or comet impacts is more likely to sustain life.

7. Potential for Prebiotic Chemistry

Prebiotic Molecules: Environments that can produce complex organic molecules (prebiotic chemistry) are crucial for the emergence of life.

Oceans and Lakes: Liquid environments like oceans or lakes could serve as "primordial soups" for the formation of prebiotic molecules.

8. Adaptability and Extremophiles

Extremophiles as a Model: Studying extremophiles on Earth helps us understand the limits of life and the adaptability of organisms to extreme environments.

Adaptability to Variable Conditions: Planetary environments that can support life adaptable to a range of conditions may widen the habitable possibilities.

As we explore diverse exoplanets, moons, and other celestial bodies in our solar system and beyond, we continuously refine our understanding of the conditions that are conducive to life. While we currently focus on life as we know it, our search for habitable environments opens the door to the possibility of discovering exotic life forms and alternative biochemistries. As technology advances and our knowledge deepens, the exploration of conditions for life on other planets takes us one step closer to unraveling the timeless question of whether we are alone in the cosmos or surrounded by a myriad of lifeforms across the universe.

The Search for Microbial Life and Intelligent Civilizations

1. The Hunt for Microbial Life

Looking Within Our Solar System: The search for microbial life begins close to home. Missions to Mars, Europa,

Enceladus, and other celestial bodies aim to explore their potential habitability and seek signs of microbial life.

Subsurface Environments: On Mars and certain moons, the possibility of life existing beneath the surface, shielded from harsh surface conditions, adds intrigue to the search.

The Viking Controversy: The Viking landers' early attempts to find signs of life on Mars yielded ambiguous results, sparking debates about the methods used and the possibility of false positives.

Extending the Search: Beyond our solar system, scientists seek to identify exoplanets in the habitable zone and study their atmospheres for potential biosignatures.

2. Biosignatures and Technosignatures

Biosignatures: These are observable indicators of biological activity that could include atmospheric imbalances (such as the coexistence of oxygen and methane), changes in the planet's surface, or unusual spectral features.

Technosignatures: These are potential signals or artifacts of advanced extraterrestrial civilizations, such as radio signals or artificial megastructures.

Listening for Intelligence: Projects like SETI (Search for Extraterrestrial Intelligence) employ radio telescopes to scan the skies for potential signals from technologically advanced civilizations.

3. The Fermi Paradox and the Great Filter

The Fermi Paradox: Coined by physicist Enrico Fermi, it highlights the apparent contradiction between the high probability of intelligent civilizations existing and the lack of evidence of their presence.

The Great Filter Hypothesis: This theory suggests that there is a critical step in the evolution of civilizations, termed the "Great Filter," that is exceptionally difficult to overcome. This could explain the scarcity of observable extraterrestrial civilizations.

4. Communication and the Challenges of Distance

Interstellar Distances: The vastness of space presents immense challenges for interstellar communication or travel, with signals taking years or even millennia to reach us.
Active and Passive SETI: Active SETI involves transmitting intentional messages into space, while passive SETI involves listening for signals without actively broadcasting.

5. The Impact of Discovery

Societal Implications: The discovery of microbial life or intelligent civilizations could profoundly impact humanity's worldview, religion, and culture.
Ethical Considerations: If we encounter extraterrestrial life, we face ethical questions about our role and responsibility in interacting with it.

6. Technological Advancements and the Future

Advancements in Telescope Technology: Upcoming telescopes like the James Webb Space Telescope (JWST) and the Square Kilometre Array (SKA) will offer unprecedented capabilities for studying exoplanets and searching for signals from intelligent civilizations.
Breakthrough Initiatives: Private initiatives like the Breakthrough Listen project aim to accelerate the search for extraterrestrial intelligence using cutting-edge technology.

The search for microbial life and intelligent civilizations continues to captivate scientists and the public alike. From our exploration of our neighboring planets to the quest for exoplanets and listening for distant signals, each step in this journey brings us closer to understanding our place in the cosmos. Whether we find simple microbial life or detect the presence of intelligent beings, the pursuit of knowledge about our cosmic neighbors fuels our curiosity and inspires us to explore the universe beyond the boundaries of our home planet. As we gaze at the stars, we remain open to the possibility of one day discovering signs of life or intelligence that will forever change our perspective on existence.

Chapter 6: UFO Phenomenon: Fact or Fiction?

The UFO phenomenon has been a subject of fascination, debate, and speculation for decades. Unidentified Flying Objects (UFOs) are objects or phenomena in the sky that defy conventional explanations. While some dismiss UFO sightings as mere hoaxes or misidentifications, others firmly believe they are evidence of extraterrestrial visitations. This chapter explores the history, investigations, and theories surrounding UFOs, separating fact from fiction.

Early UFO Sightings and Roswell Incident

Kenneth Arnold Case: The modern UFO era began in 1947 when pilot Kenneth Arnold reported seeing nine strange, saucer-like objects flying near Mount Rainier, Washington. This sighting gained widespread media attention, coining the term "flying saucer."

Roswell Incident: In July 1947, a mysterious object crashed near Roswell, New Mexico. The U.S. military initially announced it was a "flying disc," but later identified it as a weather balloon. The Roswell incident sparked conspiracy theories about a government cover-up of extraterrestrial contact.

Project Blue Book and Government Investigations

Project Blue Book: The U.S. Air Force's Project Blue Book, established in 1952, aimed to investigate UFO reports. While most cases were explained as natural or human-made phenomena, a small percentage remained unexplained.

Other Government Studies: Various countries, including France (GEPAN/SEPRA/GEIPAN), the UK (Project Condign), and Canada (Project Magnet), conducted their own UFO investigations.

UFO Sightings and Abductions

Close Encounters: UFO sightings are often classified into different types, ranging from distant sightings (Close Encounter of the First Kind) to encounters with physical effects (Close Encounter of the Second Kind) and alleged contact with occupants (Close Encounter of the Third Kind).
Alien Abductions: Some individuals claim to have been abducted by extraterrestrial beings, leading to narratives of close encounters, medical examinations, and psychological experiences.

Explanations and Skepticism

Misidentifications: Many UFO sightings turn out to be misidentifications of conventional objects, such as aircraft, weather balloons, or astronomical phenomena.
Psychological and Cultural Factors: Human psychology, imagination, and cultural influences can contribute to the perception of unusual phenomena as UFOs or alien encounters.

The Search for Extraterrestrial Intelligence (SETI)

Distinguishing UFOs from SETI: The scientific search for extraterrestrial intelligence focuses on seeking signals from advanced civilizations, while UFO sightings deal with unexplained aerial phenomena.
SETI Efforts: Organizations like SETI Institute and METI International actively search for intelligent extraterrestrial civilizations by analyzing radio signals and attempting interstellar messaging.

The Role of Pop Culture

UFOs in Media: UFOs have been a prominent theme in movies, television shows, books, and conspiracy theories, influencing public perception and beliefs about extraterrestrial life.

Hoaxes and Misinformation: The prevalence of UFO-related hoaxes and misinformation on the internet can contribute to confusion and sensationalism.

UFO Phenomenon and National Security

Declassified UFO Reports: Governments have declassified numerous UFO files, which sometimes reveal intriguing and unexplained sightings but lack definitive proof of extraterrestrial visitors.

Recent Developments: In recent years, the U.S. government has released declassified UFO reports and established a task force to investigate encounters with unidentified aerial phenomena.

Critical Thinking and the Future of UFO Research

Scientific Approach: Applying critical thinking, skepticism, and the scientific method is crucial in evaluating UFO claims objectively.

Advancing Technology: Advancements in sensor technology, astronomy, and data analysis may shed light on unidentified aerial phenomena and reduce the number of unexplained cases.

The UFO phenomenon remains a complex and polarizing topic, provoking both wonder and skepticism. While there have been numerous intriguing sightings and unexplained incidents, the search for evidence of extraterrestrial intelligence remains ongoing. Separating fact from fiction, applying scientific rigor, and remaining open to new discoveries will continue to shape the dialogue surrounding UFOs. As we explore the cosmos and our place within it, the UFO phenomenon serves as a reminder of humanity's enduring curiosity about the mysteries of the universe.

The History of UFO Sightings and Encounters

The history of UFO sightings and encounters stretches back centuries, with accounts of unidentified aerial phenomena recorded in various cultures throughout time. While many sightings have since been explained as natural or human-made phenomena, some cases remain mysterious and unexplained, fueling the ongoing fascination with UFOs.

Early UFO Accounts in History

Ancient Mythologies: Ancient civilizations, such as the Egyptians, Greeks, and Mayans, have depicted aerial phenomena in their mythologies and artwork. These could be interpreted as early UFO encounters or celestial events.
Medieval UFOs: Medieval texts and paintings occasionally depict strange flying objects that defy conventional explanations, leading to speculation about unidentified aerial phenomena.

The Modern UFO Era Begins

Kenneth Arnold Sighting (1947): The modern era of UFO sightings began in 1947 when private pilot Kenneth Arnold reported seeing nine crescent-shaped objects flying near Mount Rainier, Washington. His account gained widespread media attention, coining the term "flying saucer."
Roswell Incident (1947): The Roswell incident, where an unidentified object crashed near Roswell, New Mexico, further intensified public interest in UFOs. The initial U.S. military statement referred to a "flying disc" but was later explained as a weather balloon.

The 1950s: UFOs Gain Popularity

George Adamski's Claims: In the 1950s, George Adamski claimed to have communicated with extraterrestrial beings and documented alleged UFO sightings, popularizing the idea of contact with aliens.

Contactee Movement: The 1950s saw the emergence of a contactee movement, with individuals claiming direct communication or encounters with benevolent extraterrestrial beings.

Project Blue Book and Government Investigations

Project Blue Book (1952-1969): The U.S. Air Force's Project Blue Book aimed to investigate UFO reports. While most cases were explained as natural or human-made phenomena, a small percentage remained unidentified.

Other Government Studies: Several countries, including France, the UK, and Canada, conducted their own investigations into UFO sightings.

UFO Sightings and Pop Culture

Close Encounters: Various UFO sightings were classified into different types of "close encounters" based on proximity and interaction with the object or beings.

UFOs in Media: The popularity of UFO sightings and abduction accounts inspired numerous books, movies, and television shows, cementing the UFO phenomenon's place in pop culture.

The 1970s and Beyond

1973 UFO Wave: In 1973, a significant surge in UFO sightings occurred in the United States, known as the "UFO wave." Many sightings were eventually explained as misidentifications of celestial objects or human-made aircraft.

Betty and Barney Hill Case: In 1961, Betty and Barney Hill reported an alleged abduction experience, one of the first widely publicized abduction accounts.

Skepticism and Explanations

Natural and Human-Made Explanations: The majority of reported UFO sightings have been attributed to natural phenomena (e.g., meteors, clouds) or human-made objects (e.g., aircraft, satellites).
Psychological Factors: Human psychology, misperceptions, and optical illusions can contribute to the perception of unusual phenomena as UFOs.

Modern UFO Research and Disclosure

UFO Organizations: Independent organizations, such as MUFON (Mutual UFO Network), continue to investigate UFO sightings and encounters.
Government Disclosures: In recent years, governments, including the U.S., have released declassified UFO reports and acknowledged the existence of unidentified aerial phenomena, prompting public interest and further investigations.

The Future of UFO Research

Advancements in Technology: With advancements in sensor technology and data analysis, future UFO research may provide more clarity on unidentified aerial phenomena.
Objective Inquiry: Objective and rigorous inquiry is essential to distinguish genuine unexplained cases from misidentifications, hoaxes, or sensationalism.

As UFO sightings continue to be reported worldwide, the history of UFO encounters showcases humanity's enduring fascination with the unknown and our desire to explore the mysteries of the cosmos. While many UFO sightings can be explained by natural or human-made causes, the quest for the truth behind unidentified aerial phenomena remains an ongoing journey that sparks scientific curiosity, fuels public interest, and keeps the topic of UFOs firmly entrenched in modern culture.

Investigating Alleged Alien Abductions and Government Conspiracies

The phenomenon of alien abductions and government conspiracies has captivated the public's imagination for decades. Reports of individuals claiming to have been taken against their will by extraterrestrial beings, along with allegations of government cover-ups, have sparked intense debate and scrutiny. This section delves into the history of alien abduction claims, the methods of investigation, and the controversies surrounding government involvement.

The Emergence of Alien Abduction Accounts

Betty and Barney Hill (1961): One of the earliest and most well-known abduction accounts involves Betty and Barney Hill, who claimed to have been taken by beings from a UFO in New Hampshire. Their hypnosis sessions brought forth vivid abduction memories.

Pioneering Research: Researchers like Dr. John E. Mack and Dr. Bud Hopkins delved into the phenomenon, conducting interviews and hypnotic regression sessions with alleged abductees.

Investigative Approaches

Hypnotic Regression: Hypnosis has been used to help alleged abductees recall suppressed memories and details of their abduction experiences.

Psychological Examinations: Abductees often undergo psychological evaluations to assess their mental state and potential trauma related to the alleged encounters.

Physical Evidence: Investigators look for physical evidence such as scars, scoop marks, or implants claimed to be associated with abduction experiences.

Common Themes in Abduction Accounts

Extraterrestrial Beings: Abductees typically describe encounters with small, gray beings often referred to as "Greys," but other entities have also been reported.
Medical Examinations: Many accounts involve alleged medical procedures or examinations conducted by the beings.
Missing Time: Abductees frequently report gaps in their memory, known as "missing time," during which the abduction is said to occur.

Government Conspiracies and Cover-Ups

Government conspiracies and cover-ups related to UFOs and alien encounters have been the subject of numerous conspiracy theories and speculation. These claims suggest that governments, particularly the United States government, have concealed evidence of extraterrestrial visitations and actively worked to keep the public uninformed about the existence of UFOs and alien life. While some of these allegations are based on declassified documents and historical events, others remain largely unsubstantiated and highly controversial. Here are some key aspects of government conspiracies and cover-ups:

The Majestic 12 Documents

The Majestic 12, often abbreviated as MJ-12, refers to a purported secret committee of scientists, military officials, and government representatives allegedly established by an executive order by U.S. President Harry S. Truman in 1947. According to conspiracy theories, the MJ-12 was tasked with investigating and controlling the retrieval of crashed UFOs and their occupants. In 1984, ufologist William Moore and television producer Jaime Shandera received a package containing a set of documents that appeared to be official government papers referring to the MJ-12 committee. However, the authenticity of these documents has been heavily debated, with some experts suggesting they are elaborate hoaxes or disinformation.

The Roswell UFO Incident

One of the most famous incidents in UFO lore is the Roswell UFO incident of 1947. According to the initial press release, the U.S. military stated that they had recovered a "flying disc" that had crashed near Roswell, New Mexico. However, the next day, the military retracted the statement, claiming it was a weather balloon. This abrupt change in explanation sparked conspiracy theories about a government cover-up. Some believe that the military had indeed recovered a crashed extraterrestrial craft and that the weather balloon story was a cover story to hide the truth. The Roswell incident remains a subject of ongoing debate and investigation, with many researchers and skeptics offering alternative explanations.

Project Blue Book and Official UFO Investigations

Project Blue Book, conducted by the U.S. Air Force from 1952 to 1969, was the most well-known official government study of UFOs. Its stated purpose was to investigate and analyze UFO reports, determine if UFOs posed a threat to national security, and assess any potential technological advancements demonstrated by UFOs. While the vast majority of cases were explained as natural or human-made phenomena, a small percentage remained unidentified or inconclusive.
Conspiracy theorists argue that the government used Project Blue Book and other official UFO investigations to dismiss UFO sightings as misidentifications or to keep the public in the dark about legitimate UFO encounters. They claim that the government concealed evidence of UFOs and their advanced technologies to maintain secrecy and control over the subject.

Public Disclosure and Declassification

In recent years, there have been efforts by various governments to disclose UFO-related information to the public. For example, the U.S. government has declassified some UFO files, making them accessible to researchers and the public. While these disclosures have generated excitement among UFO enthusiasts, skeptics argue that most of the declassified files do not contain evidence of extraterrestrial encounters and are consistent with mundane explanations for UFO sightings.

Skepticism and Critical Analysis

Government conspiracies and cover-ups related to UFOs and alien encounters are met with skepticism by many researchers and officials. They argue that extraordinary claims require extraordinary evidence and that there is a lack of credible and verifiable evidence to support the existence of an elaborate government cover-up. Many UFO sightings have been convincingly explained as natural or human-made phenomena, hoaxes, or misidentifications, further undermining the conspiracy theories.

Psychological Explanations

Sleep-Related Disorders: Sleep-related phenomena, such as sleep paralysis, vivid dreams, or sleepwalking, may account for some abduction experiences.

Cultural Influence: Popular culture, media, and shared beliefs may influence the way people interpret and describe their experiences.

The Impact of Alien Abduction Claims

Coping Mechanisms: For some individuals, claiming to be an abductee may serve as a coping mechanism for dealing with traumatic experiences or psychological distress.

Cultural Significance: Alien abduction accounts reflect broader cultural interests and anxieties about the unknown and the potential for extraterrestrial life.

Government Disclosures and Public Perception

Release of UFO Files: Some governments have declassified UFO-related files, generating public interest and raising questions about government knowledge of alien encounters.
Public Opinion: The disclosure of government UFO files has both excited and polarized public opinion about the existence of extraterrestrial life and government secrecy.

The Challenge of Investigation

Objective Inquiry: Investigating abduction claims requires a balance of open-mindedness and skepticism, applying the scientific method to evaluate evidence.
Ethical Considerations: Researchers must handle abduction claims with sensitivity, considering the potential psychological impact on the alleged abductees.

The investigation of alleged alien abductions and government conspiracies remains a complex and challenging endeavor. While some cases may involve genuine experiences, the phenomenon's subjective nature and the lack of definitive evidence demand rigorous examination. As public interest in the mysteries of the cosmos persists, the pursuit of understanding the truth behind alien abduction claims and potential government involvement continues to intrigue and inspire the human imagination.

Separating truth from fiction in the realm of UFOlogy

Separating truth from fiction in the realm of UFOlogy is a challenging but crucial task. The field of UFOlogy encompasses a wide range of claims, sightings, and investigations, ranging from credible scientific research to sensationalized conspiracy theories. As we explore the UFO phenomenon, it is essential to apply critical thinking and skepticism while remaining open to genuine possibilities. Here are some key points to consider:

Examining UFO Sightings

Identifying Natural and Man-Made Phenomena: Many UFO sightings have plausible explanations, such as celestial objects (e.g., planets, meteors), aircraft, weather phenomena, or drones. Investigating these explanations can often dispel the mystery.

Anomalous Sightings: Some UFO sightings remain unexplained, but absence of an immediate explanation does not necessarily imply extraterrestrial origins. Investigating further and applying rigorous scientific methods is essential.

Evaluating Abduction Claims

Understanding Human Psychology: Alien abduction claims often involve elements of sleep paralysis, vivid dreams, or psychological trauma. Experts in psychology and psychiatry can help explore these factors in a compassionate and evidence-based manner.

Balancing Open-Mindedness and Skepticism: While it is crucial to treat abduction claimants with respect and sensitivity, it is also essential to approach their accounts with a critical eye, considering alternative explanations and potential biases.

Assessing Government Involvement

Historical Context: When exploring government conspiracies and cover-ups, it is essential to consider the historical context and evaluate claims based on verifiable evidence, not just speculation.

Declassified Information: While declassified documents may provide valuable insights, they may not always confirm the existence of extraterrestrial life. Critical analysis of these records is necessary to separate fact from speculation.

Engaging with UFO Communities

Diverse Perspectives: UFO communities can be diverse, encompassing serious researchers, enthusiasts, and those who promote fringe theories. Engaging with these communities allows for a broader understanding of viewpoints but requires discernment to separate evidence-based research from unfounded claims.

Responsible Media Consumption: Media coverage of UFO-related stories can vary widely, from balanced reporting to sensationalism. Relying on reputable sources and cross-referencing information can help to avoid misinformation.

Scientific Inquiry and UFO Studies

Legitimate UFO Research: There are credible scientists and organizations conducting UFO studies. Their focus is on applying scientific methods to investigate unexplained aerial phenomena and explore the possibility of extraterrestrial intelligence without jumping to conclusions.

Scientific Rigor: Applying scientific rigor involves peer-reviewed research, unbiased analysis, and a commitment to evidence-based conclusions. This approach helps distinguish credible UFO research from pseudoscience.

Promoting Rational Discussion

Encouraging Open Dialogue: Engaging in constructive and respectful discussions about UFOs encourages the exchange of ideas and knowledge without promoting baseless beliefs.

Avoiding Confirmation Bias: Being aware of confirmation bias is crucial in evaluating evidence. We should be open to the possibility that UFO sightings may have various explanations, including natural and terrestrial ones.

Maintaining Curiosity and Wonder

Exploring the Unknown: UFOs represent one of the many mysteries of our universe. Embracing curiosity and wonder drives scientific exploration while acknowledging the limits of our current understanding.

Remaining Open-Minded: While maintaining a critical mindset, being open to new evidence and discoveries is essential. Science has often been advanced by embracing unexpected findings.

In conclusion, separating truth from fiction in the realm of UFOlogy requires a balanced approach, combining critical thinking, scientific inquiry, and respect for diverse perspectives. By avoiding undue skepticism and baseless beliefs, we can foster a more constructive and informed discussion around UFOs, acknowledging the genuine mysteries while staying grounded in scientific principles. As we explore the possibilities of the unknown, we continue to unveil the wonders of the cosmos and our place within it.

Chapter 7: Communication with Extraterrestrial Intelligence

The search for communication with extraterrestrial intelligence (ETI) is one of the most profound scientific and philosophical endeavors undertaken by humanity. As we gaze at the vast cosmos, the question of whether we are alone or surrounded by other intelligent beings remains tantalizing. This chapter explores the history, challenges, and methods in the pursuit of establishing contact with potential extraterrestrial civilizations.

The Quest for Contact

The Drake Equation: Proposed by Dr. Frank Drake in 1961, the Drake Equation estimates the number of technologically advanced civilizations in our galaxy with which communication might be possible. It factors in variables such as the rate of star formation and the likelihood of planets capable of supporting life.

The Search for Extraterrestrial Intelligence (SETI): Since the 1960s, organizations like SETI Institute have utilized radio telescopes to scan the skies for potential signals or transmissions from advanced civilizations.

Challenges in Communication

Vast Distances: The immense distances between stars present a significant challenge in interstellar communication. Signals would take years or even centuries to travel between star systems.

The "Great Silence" or Fermi Paradox: The apparent absence of contact despite the vast number of potential civilizations raises questions about the prevalence of intelligent life or potential barriers to interstellar communication.

Messaging to Extraterrestrial Civilizations

Active Messaging: Projects like METI (Messaging Extraterrestrial Intelligence) have proposed sending intentional messages to nearby star systems in the hopes of initiating contact. Ethical considerations surround the potential impact of such messaging.

Cosmic Greetings: The Voyager Golden Record, launched in 1977 aboard the Voyager spacecraft, contains images, music, and greetings from Earth, intended as a message to any extraterrestrial intelligence that might encounter it.

The Challenge of Interpretation

Decoding Extraterrestrial Messages: If a signal is received, the challenge lies in deciphering the content and meaning of an extraterrestrial message, especially if it uses an entirely unfamiliar form of communication.

Universal Concepts: Attempts have been made to create interstellar messages using universal mathematical or scientific concepts that might be understood by any intelligent civilization.

METI and Ethical Considerations

The "Zoo Hypothesis": Some propose that extraterrestrial civilizations might be deliberately avoiding contact with humanity to allow us to develop independently, similar to a zoo keeping its distance from its inhabitants.

The Risk of Consequences: Some scientists and experts caution against actively broadcasting messages, fearing that contact with an advanced civilization might have unpredictable and potentially negative consequences.

Listening for Extraterrestrial Signals

The SETI Approach: SETI projects focus on listening for potential signals or transmissions from extraterrestrial civilizations rather than sending messages.

Breakthrough Listen: Launched in 2015, Breakthrough Listen is one of the most extensive SETI projects, surveying millions of stars in our galaxy and nearby galaxies for potential signals.

Communication Protocols and Challenges

The SETI Protocols: SETI researchers follow protocols in the event of potential signal detections to ensure responsible and coordinated responses.
Interstellar Time Delays: Communication over vast distances would involve long time delays, making real-time conversations practically impossible.

The Future of Communication with ETI

Advancements in Technology: Technological progress in telescopes, signal analysis, and data processing will continue to enhance our ability to search for and potentially communicate with extraterrestrial intelligence.
Scientific Collaboration: International collaboration and coordination in SETI efforts can maximize resources and expertise, fostering a global approach to the search for ETI.

The quest for communication with extraterrestrial intelligence is a journey that fuels our curiosity, expands our understanding of the cosmos, and challenges us to ponder our place in the universe. While contact with advanced civilizations remains uncertain, the pursuit of communication with ETI is a testament to humanity's unyielding spirit of exploration and wonder. As we continue to explore the cosmos and make advances in science and technology, we remain open to the possibility of one day receiving a cosmic signal that could forever transform our perception of the universe and our place within it.

The SETI (Search for Extraterrestrial Intelligence) Project

The Search for Extraterrestrial Intelligence (SETI) project is a scientific endeavor that aims to detect potential signals or transmissions from intelligent extraterrestrial civilizations. Since its inception in the 1960s, SETI has evolved into a global collaborative effort, combining radio astronomy, data analysis, and innovative technology to explore the possibility of intelligent life beyond Earth. This section delves into the history, methods, and impact of the SETI project.

Origins of SETI

The Pioneering Work of Frank Drake: In 1960, astronomer Dr. Frank Drake conducted the first modern SETI experiment known as Project Ozma. He used a radio telescope to search for radio signals from nearby stars, marking the beginning of the systematic search for extraterrestrial intelligence.

The Arecibo Message: In 1974, the Arecibo Observatory in Puerto Rico transmitted the Arecibo Message, a binary-encoded message depicting Earth's location and information about humanity, towards the globular star cluster Messier 13. This one-time transmission was intended as a demonstration of our capacity for interstellar communication.

Listening for Signals

Radio Telescopes: SETI researchers use powerful radio telescopes to scan the skies for narrowband or broadband signals that could indicate artificial origin.

SETI@home: Launched in 1999, SETI@home was a pioneering distributed computing project that allowed volunteers to donate their computer's processing power for SETI data analysis. It revolutionized the way massive data sets were processed, involving millions of participants worldwide.

The Challenges of SETI

Vastness of Space: The distances between stars and galaxies pose significant challenges in detecting signals from extraterrestrial civilizations. Signals might take hundreds or thousands of years to reach us.

The Cosmic "Needle in a Haystack": The radio spectrum is crowded with natural and human-made signals, making the identification of potential extraterrestrial signals challenging.

The SETI Protocols

Post-Detection Protocols: The SETI community has established protocols for what to do if a potential extraterrestrial signal is detected. These protocols include verifying the signal and notifying the international scientific community.

Coordination and Collaboration: SETI researchers work together to share data, observations, and analysis to ensure a coordinated and comprehensive search.

International SETI Efforts

SETI Institute: Based in California, the SETI Institute is one of the most prominent organizations in the field, conducting both radio and optical SETI research.

Breakthrough Listen: Launched in 2015 and funded by the Breakthrough Initiatives, Breakthrough Listen is a major SETI project aiming to survey millions of stars in our galaxy and nearby galaxies for potential signals.

SETI and Technological Advancements

Advances in Signal Processing: Improvements in signal processing and data analysis enable SETI researchers to distinguish between artificial and natural signals with greater accuracy.

Optical SETI: In addition to radio signals, some projects explore the possibility of detecting optical signals, such as laser communications, as potential methods of interstellar communication.

Public Engagement and Education

Public Support: The SETI project has garnered significant public interest and support, inspiring millions of people worldwide to engage with the scientific search for extraterrestrial intelligence.

Inspiring the Next Generation: SETI initiatives have played a role in sparking interest in science and astronomy among students and the general public.

Philosophical and Societal Implications

A Broader Perspective: The search for extraterrestrial intelligence challenges our perspective of humanity's place in the cosmos and our uniqueness as intelligent beings.

Impact on Society: The potential discovery of extraterrestrial intelligence could profoundly influence society, culture, and our understanding of existence.

The SETI project represents humanity's quest for cosmic companionship and our eagerness to explore the unknown. While no confirmed extraterrestrial signals have been detected to date, the continuous advancement of technology and international collaboration ensures that the search for intelligent life beyond Earth remains an ongoing, inspiring, and awe-inspiring endeavor. As we explore the cosmos through SETI, we recognize that even if we never find a signal, the quest itself has already expanded our knowledge and understanding of the universe, fostering curiosity and unity among the inhabitants of our pale blue dot.

Dr. Steven Greer

Dr. Steven M Greer is an American ufologist and retired physician who founded the Center for the Study of

Extraterrestrial Intelligence (CSETI) and the Disclosure Project, which seeks the disclosure of alleged classified UFO information.

Greer founded the Center for the Study of Extra-Terrestrial Intelligence (CSETI) in 1990 to create a diplomatic and research-based initiative to contact extraterrestrial civilizations. The group defined CE-5 or 'close encounters of the fifth kind' as human initiated contact and communication with extraterrestrial life. CSETI claims to have over 3,000 confirmed reports of UFO sightings by pilots and over 4,000 of what they describe as landing traces.
The organization utilizes 'Rapid Mobilization Investigative Teams' with the aim of arriving at landing sites as quickly as possible. CSETI has defined a protocol for human initiated contact to UFOs using consciousness.

In 1993, Greer founded the Disclosure Project, the goal of which is to publicly disclose the government's alleged knowledge of UFOs, extraterrestrial intelligence, and advanced energy and propulsion systems. Greer describes the Disclosure Project as an effort to grant amnesty to government whistleblowers willing to violate their security oaths by sharing classified information about UFOs.Greer claims to have briefed CIA director James Woolsey at a dinner party, although Woolsey disputes the accuracy of Greer's claim

Wikipedia contributors. (2023, July 28). Steven M. Greer. In *Wikipedia, The Free Encyclopedia.*

Efforts to Send Messages to Potential Alien Civilizations

The desire to initiate contact with potential extraterrestrial civilizations has led to various efforts to send intentional messages from Earth into the cosmos. While these attempts are not without controversy and ethical considerations, they represent humanity's proactive approach in reaching out to the unknown. This section explores some of the prominent projects and initiatives aimed at communicating with potential alien civilizations.

The Arecibo Message

In 1974, scientists at the Arecibo Observatory in Puerto Rico transmitted the Arecibo Message, an interstellar radio message aimed at the globular star cluster Messier 13, which lies approximately 25,000 light-years from Earth. The message was composed of binary-encoded data representing basic information about Earth, human beings, and our solar system. As a one-time transmission, the Arecibo Message served as a demonstration of our capability for interstellar communication.

The Pioneer Plaques

The Pioneer 10 and Pioneer 11 spacecraft, launched in 1972 and 1973, carried plaques with engraved messages. Designed by Dr. Carl Sagan and Dr. Frank Drake, the plaques depicted the silhouette of a human figure, a map of the solar system's location relative to nearby pulsars, and the hyperfine transition of hydrogen, intended as a universal unit of time and distance. The Pioneer spacecraft are now on trajectories that will take them out of the solar system, potentially serving as artifacts of human presence in interstellar space.

The Voyager Golden Record

Launched in 1977 aboard the Voyager 1 and Voyager 2 spacecraft, the Voyager Golden Record is a phonograph record containing a diverse selection of sounds, music, images, and greetings from Earth. The content was curated to showcase the diversity of life and culture on our planet. The Voyager spacecraft are now on interstellar trajectories, serving as our messengers to any potential extraterrestrial civilizations they may encounter in the distant future.

The New Horizons Message Initiative

In 2006, the New Horizons spacecraft, on its way to Pluto and beyond, carried a digital message known as the New Horizons Message Initiative. This message was crowdsourced and contained pictures, music, and text submitted by people from around the world, expressing their hopes, dreams, and visions for the future.

Messaging Extraterrestrial Intelligence (METI)

While most previous attempts were passive messages sent in the hope of eventual discovery, the METI initiative takes an active approach. METI proponents argue that rather than waiting for a signal, humanity should intentionally transmit powerful and purposeful messages to nearby stars. The objective is to initiate a two-way communication, provided that an extraterrestrial civilization is listening and responding.

Ethical Considerations and the "Zoo Hypothesis"

Sending messages to potential alien civilizations raises ethical questions and concerns. Critics of active messaging caution that we may not fully understand the implications of such actions, and the consequences of contacting an advanced and potentially unknown civilization could be unpredictable. The "Zoo Hypothesis" proposes that extraterrestrial civilizations might be deliberately avoiding contact with humanity to allow us to develop independently, similar to a zoo keeping its distance from its inhabitants.

The SETI Approach

While active messaging initiatives like METI are controversial, the majority of the scientific community involved in the search for extraterrestrial intelligence prefer the passive approach of SETI, focusing on listening for signals rather than intentionally sending messages. This approach allows us to explore the cosmos while minimizing potential risks associated with active transmissions.

In conclusion, the efforts to send messages to potential alien civilizations reflect humanity's curiosity, imagination, and yearning for cosmic companionship. Whether through passive messages on space probes or active initiatives like METI, the quest for contact with intelligent extraterrestrial beings serves as a testament to our adventurous spirit and our unyielding desire to explore the vast unknowns of the universe. As we continue our search for extraterrestrial intelligence, we do so with a sense of responsibility, ethical considerations, and a profound awareness of the potential impact such contact could have on our species and our place in the cosmos.

Ethical Considerations in Interstellar Communication

The prospect of interstellar communication with potential extraterrestrial civilizations is a fascinating and complex endeavor that raises profound ethical questions. As humanity contemplates the possibility of reaching out to the cosmos, there is an increasing awareness of the potential consequences and responsibilities that come with such actions. This section explores some of the key ethical considerations in interstellar communication and the responsible approach that should guide our interactions with the unknown.

The Precautionary Principle

The Precautionary Principle suggests that when there is a lack of scientific consensus regarding the potential risks of an action, a cautious approach should be taken to prevent harm. In the context of interstellar communication, this principle calls for careful consideration of the potential impact of contacting an unknown and potentially advanced extraterrestrial civilization.

Impact on the Recipient Civilization

Active interstellar messaging initiatives, such as METI, raise concerns about the impact on the recipient civilization. If an extraterrestrial civilization exists and receives our message, how might they interpret and react to our communication? There is no guarantee that they would have the same intentions or values as we do, potentially leading to unintended consequences.

Unpredictable Outcomes

The consequences of interstellar communication are uncertain and unpredictable. It is challenging to foresee how an advanced civilization might react to our message or what long-term effects our communication could have on both our species and theirs.

Cultural and Linguistic Challenges

Communicating with an extraterrestrial civilization would present significant cultural and linguistic challenges. The vast differences in technological development, societal structures, and modes of communication could hinder meaningful understanding between the two civilizations.

Preservation of Earth's Identity

Sending messages into space that represent Earth and humanity requires consideration of the values, beliefs, and diversity of our planet. Ensuring that any messages we send accurately reflect our shared human experience while being respectful of various cultural perspectives is crucial.

Potential Risks to Earth

Active messaging initiatives carry the theoretical risk of attracting the attention of advanced civilizations that may not have our best interests in mind. Critics argue that intentionally broadcasting our presence could invite unknown consequences to Earth.

Long-Term Impact

Interstellar messages have the potential to persist in space indefinitely. Consideration must be given to the long-term implications of our actions, as the consequences of interstellar communication could extend far beyond our current time and understanding.

Informed Consent of Humanity

Active messaging initiatives raise questions about informed consent. Advocates argue that a decision to transmit messages should be made collectively by the entire human population, as it impacts the future of all humanity.

International Cooperation and Governance

Interstellar communication is a matter that concerns all of humanity. Establishing international cooperation and governance frameworks to address the ethical, cultural, and societal aspects of potential communication with extraterrestrial civilizations is essential.

The Significance of Contemplating the Unknown

Engaging in ethical considerations surrounding interstellar communication is a reflection of our capacity for thoughtful reflection, empathy, and respect for the unknown. This process allows us to think critically about our place in the cosmos and how our actions may reverberate beyond our planet.

In conclusion, ethical considerations play a central role in the quest for interstellar communication. As humanity explores the possibility of reaching out to other civilizations, a responsible and cautious approach is vital. Balancing our curiosity and ambition with a sense of humility and respect for the unknown is essential in shaping the future of our interstellar endeavors. The pursuit of interstellar communication should be guided by a collective effort to safeguard humanity's interests and preserve the beauty, diversity, and values of our home planet.

Chapter 8: First Contact: Fictional Scenarios

The concept of "First Contact" with extraterrestrial civilizations has captured the human imagination for centuries. In this chapter, we delve into fictional scenarios that explore the intriguing possibilities and challenges surrounding the first encounter with intelligent beings from beyond Earth. These stories, from literature, movies, and television, have shaped our perceptions and emotions regarding the profound moment of first contact.

Literature: "Contact" by Carl Sagan

Carl Sagan's novel "Contact," later adapted into a film, tells the story of Ellie Arroway, a SETI scientist who detects a mysterious signal from the star Vega. The contact leads to the construction of a complex interstellar travel machine based on alien blueprints, eventually taking Ellie on a breathtaking journey of discovery. The novel explores themes of faith, scientific inquiry, and the awe-inspiring idea of communication across the vast cosmic distances.

Film: "Close Encounters of the Third Kind"

Directed by Steven Spielberg, "Close Encounters of the Third Kind" is a classic science fiction film that revolves around the arrival of an alien spacecraft and the government's efforts to communicate with the extraterrestrial visitors. The film portrays both the wonder and fear that humans experience in the face of the unknown, while emphasizing the importance of peaceful communication and understanding.

Television: "Star Trek: First Contact"

In this "Star Trek: The Next Generation" film, the crew of the starship Enterprise travels back in time to Earth's past, aiming to prevent a pivotal moment in history from being altered by hostile cybernetic beings known as the Borg. "First Contact" highlights the challenges of interacting with less advanced civilizations while adhering to the principle of non-interference.

Literature: "The War of the Worlds" by H.G. Wells

H.G. Wells' iconic novel "The War of the Worlds" depicts a hostile invasion of Earth by Martians. The story explores humanity's struggle for survival and our vulnerability when confronted with technologically superior beings. The novel also reflects on the impact of colonialism, as the Martians treat humans as we have treated other species on Earth.

Film: "Arrival"

In "Arrival," directed by Denis Villeneuve and based on Ted Chiang's novella "Story of Your Life," linguist Louise Banks is tasked with deciphering the language of enigmatic extraterrestrial visitors known as Heptapods. The film delves into the complexities of language, time, and the nature of perception, as Louise's understanding of the alien language transforms her perception of reality.

Television: "Doctor Who"

The long-running British television series "Doctor Who" frequently explores the theme of first contact with alien civilizations. The Doctor, a time-traveling alien himself, often finds himself mediating interactions between humans and extraterrestrial beings, advocating for understanding and empathy.

Literature: "Childhood's End" by Arthur C. Clarke

In this novel by Arthur C. Clarke, highly advanced and benevolent extraterrestrial beings known as the Overlords arrive on Earth, ushering in a new era of peace and prosperity. However, as humanity begins to evolve, the Overlords' true intentions and the cost of their intervention become apparent, raising thought-provoking questions about the nature of progress and the role of higher civilizations in guiding others.

Film: "Independence Day"

In "Independence Day," directed by Roland Emmerich, Earth faces a sudden and hostile invasion by technologically superior extraterrestrial forces. The film portrays humanity's fight for survival and the unifying spirit that arises when faced with a common threat.

Television: "The X-Files"

"The X-Files" television series revolves around FBI agents Mulder and Scully as they investigate paranormal and unexplained phenomena, including encounters with extraterrestrial beings. The show captures the intrigue and mystery surrounding the idea of hidden truths and government cover-ups related to alien contact.

These fictional scenarios provide diverse perspectives on the potential outcomes and ramifications of first contact with extraterrestrial civilizations. They reflect our hopes, fears, and aspirations about our place in the universe and the mysteries that lie beyond our understanding. Whether filled with wonder, danger, or philosophical musings, these stories remind us of the significance of the search for intelligent life and the profound impact such a discovery could have on humanity's collective journey through the cosmos.

Imagining different scenarios for human-alien interaction

Imagining different scenarios for human-alien interaction allows us to explore the vast array of possibilities that may arise if we ever encounter intelligent extraterrestrial beings. These scenarios, ranging from peaceful cooperation to tense conflicts, capture the diversity of human imagination and reflect our hopes and anxieties about the unknown. Let's delve into some of these intriguing scenarios:

1. Peaceful Diplomacy and Cooperation

In this scenario, humanity and extraterrestrial beings meet with mutual respect and a desire to learn from each other. Peaceful diplomacy is established, leading to cultural and technological exchanges that benefit both civilizations. Collaborative efforts focus on solving shared challenges, such as environmental issues, medical advancements, and scientific discoveries. The encounter fosters a new era of interstellar cooperation, opening the door to a harmonious coexistence in the cosmos.

2. Scientific Curiosity and Cultural Exchange

Human-alien interaction becomes a scientific venture where both civilizations seek to understand each other's biology, technology, and culture. Scientists collaborate on studying each other's languages, social structures, and histories, leading to groundbreaking discoveries about the nature of intelligent life in the universe. The exchange of knowledge and ideas enriches both societies and ignites a passion for interstellar exploration and learning.

3. Interstellar Conflict and Hostility

This scenario portrays a more somber encounter, where misunderstandings and misinterpretations fuel tensions between humanity and extraterrestrial beings. Fear, suspicion, and different value systems create a potential for conflict. Both sides might resort to defensive measures, leading to confrontations and skirmishes. The scenario serves as a cautionary tale, urging us to approach the unknown with care and empathy.

4. Covert Observation and Non-Interference

Extraterrestrial beings observe humanity from a distance, practicing a policy of non-interference, similar to the "Prime Directive" in "Star Trek." They remain hidden, allowing humanity to develop on its own without direct interference. The scenario raises philosophical questions about the responsibility of advanced civilizations in the cosmos and their role in guiding less developed ones.

5. An Unveiling of Ancient Knowledge

In this scenario, extraterrestrial beings reveal themselves as ancient custodians of knowledge, visiting Earth periodically throughout history. They share their wisdom, guiding humanity's progress from behind the scenes. The revelation of this cosmic heritage triggers profound shifts in human understanding and spirituality.

6. Parallel Universes and Alternate Realities

Human-alien interaction transcends traditional space-time boundaries, leading to encounters with beings from parallel universes or alternate realities. The scenario challenges our understanding of reality and stretches the boundaries of what is possible.

7. Unintelligible Communication and Linguistic Challenges

In this scenario, communication barriers pose significant challenges. Both civilizations struggle to comprehend each other's languages or methods of communication. Scientists and linguists work tirelessly to decipher alien languages, leading to innovative approaches in linguistics and the discovery of new ways to convey meaning.

8. A Revelation of Cosmic Purpose

Extraterrestrial beings arrive to reveal a grand cosmic purpose that unites all intelligent life in the universe. Their message transcends cultural and societal differences, offering profound insights into the interconnectedness of all life forms.

These scenarios represent a mere fraction of the myriad possibilities for human-alien interaction. As we continue to search for extraterrestrial intelligence and ponder the mysteries of the cosmos, these imaginative scenarios remind us of the boundless potential for discovery and exploration. Each scenario encourages us to contemplate our place in the universe, the significance of our actions, and the impact of potential encounters with intelligent beings beyond Earth.

Classic and contemporary portrayals of aliens in literature and movies

Classic Portrayals of Aliens in Literature and Movies
"The War of the Worlds" (1898) by H.G. Wells: Wells' novel is a seminal work in alien invasion literature. The Martians arrive in large, tripod-like machines, launching a devastating assault on Earth. The story explores themes of colonialism, human vulnerability, and the consequences of meeting technologically advanced beings.
"E.T. the Extra-Terrestrial" (1982): Directed by Steven Spielberg, this heartwarming film portrays a friendly alien stranded on Earth who befriends a young boy. E.T. explores themes of friendship, empathy, and the wonder of discovering life beyond our planet.
"The Day the Earth Stood Still" (1951): In this classic science fiction film, an alien named Klaatu arrives on Earth with a

powerful robot named Gort. Klaatu brings a message of peace but faces hostility from humanity. The film explores themes of nuclear warfare and the potential consequences of our actions as a species.

"Solaris" (1961) by Stanislaw Lem: This philosophical science fiction novel tells the story of a sentient ocean on a distant planet that manifests visitors' memories and desires. The encounter with this enigmatic extraterrestrial entity raises profound questions about consciousness and the nature of reality.

Contemporary Portrayals of Aliens in Literature and Movies

"Arrival" (2016): Directed by Denis Villeneuve, this film explores the arrival of enigmatic extraterrestrial beings known as Heptapods. A linguist, played by Amy Adams, must decipher their complex language to understand their intentions and the implications of their presence.

"The Three-Body Problem" (2008) by Liu Cixin: This award-winning Chinese science fiction novel explores humanity's communication with an alien civilization, the Trisolarans. The novel delves into themes of physics, cultural differences, and the challenges of interstellar contact.

"District 9" (2009): Directed by Neill Blomkamp, this gritty film portrays an alien refugee population stranded in South Africa. The movie serves as an allegory for issues such as xenophobia, segregation, and the consequences of human exploitation.

"The Expanse" (book series by James S.A. Corey): This science fiction series, adapted into a popular TV show, depicts humanity's encounter with an ancient alien protomolecule that grants immense power but poses a significant threat. The story explores political intrigue, interstellar conflict, and the potential consequences of meddling with advanced alien technology.

"Contact" (1997): Based on Carl Sagan's novel, the film follows a scientist's efforts to decode a mysterious message from extraterrestrial beings. It explores themes of faith, scientific discovery, and the implications of first contact.

Contemporary portrayals of aliens in literature and movies often delve into complex themes of human nature, the search for meaning, and the ethical dilemmas that arise from potential encounters with intelligent beings from beyond Earth. These stories reflect our evolving understanding of science, technology, and the cosmos while continuing to captivate our imaginations with the profound mysteries of the universe.

The impact of first contact on human society and culture

The impact of first contact with extraterrestrial civilizations would undoubtedly be one of the most profound and transformative events in human history. Such an encounter would have far-reaching implications, touching on nearly every aspect of human society and culture. Here are some of the ways that first contact could reshape humanity's worldview and way of life:

1. Paradigm Shift in Worldview

First contact would challenge fundamental aspects of human beliefs and understanding. The knowledge that we are not alone in the universe would likely lead to a reevaluation of our place in the cosmos, questioning our significance and uniqueness. Religious and philosophical frameworks might undergo revision as new questions arise about the nature of life, creation, and the divine.

2. Cultural Exchange and Diversity

The exchange of knowledge and ideas with extraterrestrial civilizations could lead to an unprecedented cultural exchange. Different belief systems, arts, languages, and social structures from alien worlds would enrich humanity's cultural tapestry, fostering a deeper appreciation for diversity and the interconnectedness of all intelligent life.

3. Technological Advancements

The introduction of advanced extraterrestrial technology could revolutionize human civilization. Scientific breakthroughs in fields such as energy, medicine, transportation, and communication might rapidly accelerate, leading to advancements that were previously unimaginable.

4. Global Collaboration

First contact would likely bring humanity together as a species. The realization that we share a cosmic neighborhood with other intelligent beings might encourage global cooperation and unity in addressing common challenges and goals, such as environmental preservation and space exploration.

5. Ethical Considerations

Encountering an alien civilization raises profound ethical questions. Debates would arise about our responsibilities as a species, the implications of interacting with less advanced civilizations, and the potential consequences of our actions on both Earth and alien worlds.

6. New Frontiers of Exploration

The prospect of first contact could reinvigorate humanity's enthusiasm for space exploration. The desire to learn more about the cosmos and seek out other intelligent life forms might drive us to explore distant star systems and set up interstellar missions.

7. Social Impact

First contact would trigger a range of emotional responses, including awe, fear, curiosity, and excitement. Governments and institutions would face the challenge of managing public reactions, ensuring transparency, and addressing concerns while providing reassurance.

8. Reevaluation of Conflict and Cooperation

The existence of extraterrestrial civilizations might prompt humans to reconsider intergroup conflicts on Earth. The encounter with an entirely different species could serve as a powerful reminder of our shared humanity and the need for peaceful cooperation.

9. Environmental Awareness

The realization that Earth is just one of many inhabited planets might strengthen humanity's commitment to environmental stewardship. Understanding the fragility of life in the cosmos could deepen our understanding of the interconnectedness of all living beings.

10. Redefining Identity and Purpose

First contact could lead to a reevaluation of human identity and purpose. The encounter might inspire us to strive for higher ideals and consider our role in shaping the future of the universe.

In conclusion, the impact of first contact on human society and culture would be multi-faceted and profound. Such an event would challenge our beliefs, reshape our worldview, and foster a sense of interconnectedness with other intelligent beings in the cosmos. The encounter would present both opportunities and challenges, requiring humanity to navigate uncharted territories with wisdom, humility, and a commitment to global cooperation. Ultimately, first contact would be a transformative moment that shapes the course of human history and propels us into a new era of interstellar exploration and understanding.

Chapter 9: The Fermi Paradox: Where Are They?

The Fermi Paradox is a perplexing question that arises from the apparent contradiction between the high probability of extraterrestrial civilizations existing in the vast universe and the lack of evidence or contact with them. In this chapter, we delve into the enigma of the Fermi Paradox and explore various hypotheses and possible explanations for the absence of observable extraterrestrial intelligence.

The Great Silence: The Paradox Unveiled

The Basics of the Fermi Paradox: The Fermi Paradox is named after physicist Enrico Fermi, who famously asked, "Where is everybody?" The paradox centers on the fact that, given the vast number of potentially habitable planets in the universe and the potential for advanced civilizations to have developed, we have not yet detected any signs of their existence.

The Drake Equation: The Drake Equation is a formula devised by astronomer Frank Drake to estimate the number of technological civilizations in our galaxy with whom communication might be possible. Despite the uncertainties in the variables, the equation suggests that there should be a considerable number of civilizations in the Milky Way alone.

Possible Explanations for the Fermi Paradox

The Rare Earth Hypothesis: Some propose that the emergence of intelligent life requires an exceedingly rare combination of factors, making advanced civilizations exceptionally rare in the universe. Factors such as the right type of star, a stable planetary system, and a protective atmosphere might be crucial prerequisites for complex life.

The Great Filter: The Great Filter is a hypothetical concept that suggests there could be one or more extremely improbable steps in the evolution of life that act as a "filter" to prevent the emergence of advanced civilizations. This filter could be behind us (suggesting we are rare) or ahead (suggesting the extinction of civilizations is common).

The Zoo Hypothesis: The Zoo Hypothesis proposes that extraterrestrial civilizations are deliberately avoiding contact with Earth, either to let us develop independently or to preserve our natural evolution without interference.

Technological Self-Destruction: Some theories suggest that intelligent civilizations might reach a point where they develop advanced technology capable of self-destruction, leading to their downfall before they can explore the cosmos or make contact.

Interstellar Communication Challenges: It's possible that advanced civilizations do exist, but the vast distances between stars make interstellar communication exceedingly difficult or impractical with current technology.

Civilization Lifespans: The lifespan of an advanced civilization might be relatively short in cosmic terms, making it challenging for civilizations to overlap in time and make contact with one another.

Transcendence to a Different State: Advanced civilizations might undergo a transformation or transcendence to a different state of existence, beyond our current understanding, making their presence difficult to detect.

The Future of the Fermi Paradox

The Fermi Paradox continues to inspire scientific research, philosophical debates, and the exploration of the cosmos. As we develop more advanced technologies and conduct further surveys of exoplanets, our understanding of the paradox may evolve. The possibility of detecting signs of extraterrestrial life, even microbial, in our own solar system (e.g., Mars or Europa) adds another layer of intrigue to the Fermi Paradox.

In the quest to unravel the mystery of the Fermi Paradox, scientists, astronomers, and thinkers are driven by the curiosity to explore our place in the universe and understand the potential cosmic companions we may have. As we contemplate the possibility of other intelligent beings in the cosmos, the Fermi Paradox serves as a reminder of the vastness of the universe and the profound questions that lie beyond our current knowledge.

The Fermi Paradox and Its Implications for the Existence of Extraterrestrial Civilizations

The Fermi Paradox poses one of the most compelling and enduring questions in the field of astrobiology and the search for extraterrestrial intelligence. The absence of observable extraterrestrial civilizations despite the vast number of potential habitable planets in the universe has led to a range of implications and possible scenarios. Here, we explore some of the key implications of the Fermi Paradox for the existence of extraterrestrial civilizations:

1. The Rare Earth Hypothesis Revisited

The Fermi Paradox challenges the optimistic assumptions of the Drake Equation, which suggests a considerable number of advanced civilizations in our galaxy alone. The absence of evidence for such civilizations has reignited interest in the Rare Earth Hypothesis, which proposes that the emergence of complex life, let alone intelligent civilizations, might be extremely rare in the universe. This notion suggests that Earth and humanity could be unique or nearly unique in the cosmos.

2. The Great Filter as a Barrier to Civilization

The concept of the Great Filter gains significance in the context of the Fermi Paradox. If there exists one or more improbable steps in the evolution of life that act as a filter, preventing the emergence of advanced civilizations, it would explain the lack of observable extraterrestrial intelligence. This scenario raises important questions about the potential challenges and threats that could be responsible for filtering out most civilizations before they reach the point of interstellar communication.

3. Interstellar Communication Challenges and the Speed of Light

One implication of the Fermi Paradox is that interstellar communication may be much more challenging and time-consuming than previously imagined. The vast distances between stars, combined with the limitation imposed by the speed of light, mean that even if extraterrestrial civilizations exist, the time it takes for their signals to reach us or vice versa could be prohibitive.

4. The Nature of Technological Advancements

The Fermi Paradox raises questions about the long-term stability of technological civilizations. If advanced civilizations tend to self-destruct due to the consequences of their technological advancements, it could be a cautionary tale for humanity. Examining how civilizations navigate the challenges of their own technological development becomes essential to avoid similar pitfalls.

5. Searching for Other Technological Signatures

As the search for extraterrestrial intelligence continues, scientists are expanding their scope beyond traditional radio signals. Efforts now include the search for technosignatures—indicators of advanced technology, such as megastructures, Dyson spheres, or artificial planets. These endeavors might offer new avenues for detection and increase the chances of finding evidence of intelligent civilizations.

6. The Impact on Humanity's Self-Perception

The Fermi Paradox has the potential to significantly impact humanity's self-perception and sense of uniqueness. If we were to discover evidence of extraterrestrial civilizations, it would revolutionize our understanding of our place in the universe. Conversely, if we continue to find no evidence of advanced civilizations, it could foster a sense of cosmic loneliness and isolation.

7. The Future of the Search for Extraterrestrial Intelligence

The Fermi Paradox is a driving force behind the ongoing search for extraterrestrial intelligence. As we explore exoplanets, develop more advanced technologies, and expand our knowledge of astrobiology, our understanding of the implications of the paradox may evolve. Regardless of the outcome, the quest to address the Fermi Paradox has profound implications for our understanding of the cosmos and our place within it.

In conclusion, the Fermi Paradox challenges our assumptions about the prevalence of extraterrestrial civilizations and forces us to reevaluate our understanding of the universe and the potential for intelligent life beyond Earth. The absence of observable extraterrestrial intelligence raises profound questions about the nature of life, the evolution of civilizations, and the future of humanity's exploration of the cosmos. As we continue to search for answers, the Fermi Paradox remains a captivating mystery that ignites our imagination and curiosity about the vast unknowns of the universe.

Proposed Solutions and Theories to Explain the Apparent Absence of Alien Contact

The Fermi Paradox has sparked numerous proposed solutions and theories from scientists, thinkers, and researchers attempting to explain the apparent absence of alien contact despite the high probability of extraterrestrial civilizations in the universe. These hypotheses range from cosmic filters to the limitations of our technology and perceptions. Let's explore some of the most prominent proposed solutions:

1. The Great Filter Hypothesis

One of the most intriguing theories suggests the existence of a "Great Filter" that acts as a barrier preventing the emergence of advanced civilizations. This filter could be a rare event that significantly reduces the number of civilizations capable of interstellar communication. The filter might lie in the early stages of life's emergence, complex multicellular life, the development of technology, or avoiding self-destruction. If the Great Filter lies ahead of us, it could imply that civilizations tend to destroy themselves before becoming interstellar.

2. The Technological Stagnation Theory

This theory suggests that civilizations reach a certain level of technological advancement and then enter a state of stagnation. This could be due to social, economic, or political factors that inhibit further progress or exploration beyond their home planet. As a result, advanced civilizations might remain isolated within their own star systems.

3. The Simulation Hypothesis

This hypothesis proposes that we may be living in a simulated reality created by a more advanced civilization. In such a scenario, the creators of the simulation may choose to withhold the appearance of extraterrestrial civilizations to preserve the integrity of the simulation or prevent interference with its inhabitants.

4. Interstellar Communication Difficulties

Extraterrestrial civilizations might exist, but communication between them and us is limited by the vast distances and the speed of light. Our radio signals may not have reached them yet, or their signals might not have reached us due to similar limitations. Additionally, advanced civilizations could be using communication methods beyond our current technological capabilities.

5. The Zoo Hypothesis

This hypothesis suggests that advanced extraterrestrial civilizations are aware of us but have chosen not to make contact, analogous to humans observing wildlife in a zoo without interfering. This could be out of respect for our development, the desire to observe natural evolution, or ethical concerns about influencing a less advanced civilization.

6. Cultural and Societal Differences

Advanced civilizations might operate in ways that are fundamentally different from ours, making their presence or actions challenging to detect or comprehend. Their values, technology, and communication methods might be vastly dissimilar to what we expect, making interaction and recognition challenging.

7. The Rare Intelligence Theory

While the universe may be teeming with life, the emergence of intelligent, technologically capable civilizations could be exceedingly rare. The development of higher cognitive abilities and complex societies might require an extraordinary set of circumstances that rarely occur.

8. Extraterrestrial Hibernation

It's possible that some advanced civilizations, for reasons unknown, choose to hibernate or remain dormant for extended periods, making them temporarily invisible to us.

In conclusion, the apparent absence of alien contact is a fascinating and enduring mystery that continues to captivate scientists and the public alike. The proposed solutions and theories to explain the Fermi Paradox provide a diverse range of possibilities, ranging from technological limitations to profound cosmic filters. As we explore the cosmos and develop new technologies, our understanding of this paradox may evolve, leading us closer to unraveling the secrets of potential extraterrestrial civilizations and our place in the universe.

Chapter 10: The Future of Space Exploration and Alien Contact

The search for extraterrestrial life and the quest for alien contact have captured the imagination of humanity for generations. As our understanding of the cosmos deepens and technology advances, the future of space exploration holds exciting possibilities and potential encounters with intelligent beings beyond Earth. In this chapter, we explore the potential directions of space exploration and the implications of possible alien contact.

Advancements in Space Exploration:

Exoplanet Studies: As our technology improves, we continue to discover exoplanets within the habitable zones of their stars, where liquid water might exist. Space missions and telescopes like the James Webb Space Telescope and the next-generation ground-based telescopes will enable us to study these exoplanets in greater detail, searching for biosignatures and potential signs of life.

Interstellar Probes: Advancements in propulsion technology may lead to the development of interstellar probes capable of traveling to neighboring star systems. These probes could carry scientific instruments to study exoplanets up close and transmit data back to Earth.

SETI and Technosignature Searches: The Search for Extraterrestrial Intelligence (SETI) continues to scan the skies for artificial signals that might indicate the presence of advanced civilizations. The search for technosignatures, such as signs of advanced technology or engineering, complements traditional SETI efforts.

Astrobiology Research: Astrobiologists study extremophiles on Earth to understand the conditions under which life can thrive. This knowledge informs our search for potentially habitable environments beyond Earth, such as subsurface oceans on icy moons like Europa and Enceladus.

Preparing for Alien Contact:

Ethical and Societal Considerations: As we approach the possibility of contact with intelligent extraterrestrial beings, we must grapple with ethical dilemmas and societal implications. Establishing protocols for communication and coordination among nations becomes crucial to navigate potential encounters respectfully and responsibly.

International Collaboration: Alien contact would be a momentous event for all of humanity. International collaboration and coordination will be vital to ensure that information is shared transparently and that decisions are made collectively.

Cultural and Philosophical Preparedness: Societies around the world may need to adapt to the paradigm-shifting implications of alien contact. Preparing for potential changes in belief systems, societal structures, and cultural norms becomes essential.

Possible Outcomes of Alien Contact:

Scientific Discoveries: Alien contact would provide an unprecedented opportunity for scientific advancement. The exchange of knowledge and ideas with extraterrestrial civilizations could lead to breakthroughs in physics, biology, and other fields.

Technological Insights: Interaction with advanced civilizations might offer insights into cutting-edge technologies that could revolutionize various industries on Earth.

Cosmic Cooperation: Contact with extraterrestrial civilizations could unite humanity in a common goal of interstellar exploration and understanding the universe.

Societal Reflection and Transformation: Alien contact might prompt humanity to reevaluate our place in the cosmos, foster a sense of interconnectedness, and inspire us to address global challenges collectively.

The Ongoing Search:

While the future of space exploration and the possibility of alien contact are filled with excitement, we must remember that they are explorations of the unknown. The search for extraterrestrial life is a process of patient and meticulous scientific investigation. As we venture further into the cosmos, we may encounter phenomena that defy our current understanding, leading to more questions and mysteries.

In conclusion, the future of space exploration and the potential for alien contact are among the most captivating endeavors of humanity. Advancements in technology, international cooperation, and ethical considerations will shape our journey into the cosmos. Whether we make contact in the near future or continue to explore the cosmos with wonder and curiosity, the pursuit of understanding our place in the universe will continue to inspire and push the boundaries of human knowledge.

The Future of Space Exploration and the Role of Private Space Companies

The future of space exploration is entering an exciting phase with the emergence of private space companies. These companies are driving innovation, advancing technology, and expanding humanity's reach into the cosmos. In this chapter, we explore the role of private space companies and their impact on the future of space exploration.

Advancements in Technology:

Reusable Rockets: Private space companies, such as SpaceX with its Falcon 9 and Falcon Heavy rockets, have pioneered the development of reusable rocket technology. This breakthrough dramatically reduces the cost of space travel and opens up new possibilities for exploring the cosmos.

Commercial Space Travel: Companies like Blue Origin and Virgin Galactic are working towards making space travel accessible to private individuals. Suborbital and orbital space tourism could become a reality, allowing more people to experience the wonders of space.

Space Mining: Private companies are exploring the potential for mining valuable resources from asteroids and other celestial bodies. Space mining could revolutionize resource availability on Earth and support future space missions.

Expanding Human Presence:

Lunar Exploration: Private companies are partnering with space agencies to return humans to the Moon. These collaborations aim to establish lunar habitats, conduct scientific research, and prepare for deeper space exploration.

Mars Colonization: SpaceX's ambitious goal of colonizing Mars has captured global attention. Private companies are actively developing the technologies needed for long-duration missions to the Red Planet, laying the groundwork for potential human settlement.

Space Stations and Habitats: Private companies are investing in the development of commercial space stations and habitats in low Earth orbit. These facilities could serve as research laboratories, manufacturing centers, and even tourist destinations.

Collaboration with Space Agencies:

Public-Private Partnerships: Private space companies are collaborating with space agencies like NASA and ESA to enhance space capabilities. These partnerships leverage the expertise of both sectors to advance space exploration.

Resupply and Crew Transport: Private companies are providing cargo resupply and crew transportation services to the International Space Station (ISS). These services ensure a continuous human presence in space and support scientific research.

Pushing Boundaries and Inspiring the Next Generation:

Space Tourism: Private space companies are on the cusp of making space tourism a reality. Space tourists and adventurers will have the opportunity to experience weightlessness and see the curvature of the Earth, inspiring a new generation of space enthusiasts.

Innovation and Competition: The rise of private space companies has sparked a new era of innovation and healthy competition in the space industry. This competition drives progress, accelerates technological advancements, and lowers the cost of space missions.

Addressing Global Challenges:

Satellite Constellations: Some private companies are deploying large constellations of satellites to provide global internet coverage. These initiatives have the potential to bridge the digital divide and provide connectivity to underserved regions.

Earth Observation: Private space companies are launching sophisticated Earth observation satellites to monitor and address environmental challenges, such as climate change, deforestation, and natural disasters.

In conclusion, the future of space exploration is increasingly intertwined with the efforts of private space companies. Their innovations, ambitious goals, and collaborations with space agencies are reshaping humanity's understanding of the cosmos. As private space companies continue to push the boundaries of technology and explore new frontiers, they inspire a renewed sense of wonder, curiosity, and excitement about our place in the universe. Through their endeavors, these companies play a vital role in expanding human presence beyond Earth and paving the way for the next phase of space exploration.

Technological advancements that may aid in alien contact

Technological advancements play a critical role in our ongoing quest for alien contact and the search for extraterrestrial intelligence (SETI). As our technology improves, we enhance our ability to detect, communicate with, and understand potential extraterrestrial civilizations. Here are some technological advancements that may aid in alien contact:

Advancements in Radio Telescopes:

Radio telescopes have been the primary tool in the search for alien signals. Technological improvements in radio telescopes, such as larger apertures, increased sensitivity, and more sophisticated data analysis algorithms, have allowed us to scan the skies with greater precision and coverage. Future radio telescopes, like the Square Kilometre Array (SKA), will significantly enhance our ability to detect weak signals from potential alien civilizations.

Optical and Infrared Telescopes:

Optical and infrared telescopes are vital in the search for exoplanets and the study of their atmospheres for biosignatures. Technological advancements in adaptive optics and coronagraphs have enabled us to directly image exoplanets and gather critical data about their compositions. The James Webb Space Telescope (JWST) is expected to revolutionize exoplanet studies and pave the way for future space-based observatories.

High-Performance Computing:

Analyzing the vast amount of data collected from telescopes and space missions requires immense computational power. High-performance computing allows us to process and analyze large datasets, identifying potential candidate signals and distinguishing them from natural or human-made interference.

Laser Communication:

Traditional radio waves may not be the most efficient means of interstellar communication due to their dispersion and limited bandwidth. Laser communication, using directed laser beams, could provide a more focused and efficient way to send and receive messages across vast distances. This technology could improve the feasibility of interstellar communication if we encounter advanced civilizations capable of using laser-based communication systems.

Interstellar Space Probes:

Advancements in propulsion technology, such as nuclear or solar sails, could enable the development of interstellar space probes capable of reaching nearby star systems within a reasonable time frame. These probes could carry encoded messages or information about humanity and Earth, serving as emissaries for potential interstellar communication.

Machine Learning and Artificial Intelligence:

Machine learning and artificial intelligence (AI) are revolutionizing various scientific fields, including astronomy and SETI. AI algorithms can help sift through vast datasets, identify anomalies, and classify potential candidate signals that might indicate extraterrestrial intelligence.

Decentralized and Distributed Networks:

To improve the global search for alien signals, decentralized and distributed networks of telescopes and radio antennas can collaborate to achieve a wider coverage of the sky. These networks can share data in real-time, allowing for coordinated observations and rapid detection of interesting signals.

Deep Space Communication Infrastructure:

Establishing robust and efficient deep space communication networks is crucial for exchanging messages with potential extraterrestrial civilizations. Advancements in space communication technology, such as optical communication links, can significantly enhance our ability to communicate over interstellar distances.

In conclusion, technological advancements are fundamental in our pursuit of alien contact and understanding the potential existence of extraterrestrial civilizations. As our technology evolves, we continue to expand our knowledge of the cosmos and improve our ability to detect, communicate with, and potentially interact with intelligent beings beyond Earth. The ongoing development of cutting-edge technologies in various fields will undoubtedly play a pivotal role in shaping the future of space exploration and our search for extraterrestrial life.

Humanity's readiness for the possibility of encountering extraterrestrial life

Humanity's readiness for the possibility of encountering extraterrestrial life is a complex and multifaceted issue that goes beyond scientific and technological preparedness. It encompasses societal, cultural, ethical, and philosophical considerations. Here are some key aspects to consider regarding humanity's readiness for such a momentous event:

Scientific Preparedness:

Scientific institutions and organizations play a crucial role in preparing for the possibility of encountering extraterrestrial life. Continued research in astrobiology, exoplanet studies, and the search for technosignatures helps us better understand the conditions for life in the universe and enhances our ability to detect potential signs of extraterrestrial intelligence.

International Collaboration:

The search for extraterrestrial life is a global endeavor. International collaboration among nations and space agencies fosters cooperation, resource-sharing, and the exchange of knowledge. An open and transparent exchange of data and findings ensures that the entire world can benefit from any significant discoveries.

Ethical and Philosophical Reflection:

Encountering extraterrestrial life, especially intelligent civilizations, raises profound ethical and philosophical questions. Societies must engage in thoughtful reflection and dialogue about the potential consequences of such contact. This includes discussions about the ethical treatment of alien life forms, respecting their autonomy, and avoiding harmful interference.

Cultural and Religious Sensitivity:

Different cultures and belief systems have diverse perspectives on the existence of extraterrestrial life and the implications of contact. A respectful and inclusive approach that acknowledges cultural diversity can foster a constructive dialogue about the potential impact of such an encounter on various societies.

Media and Public Perception:

The media's role in shaping public perception is significant. Responsible and accurate reporting on the topic of extraterrestrial life can help manage public expectations and avoid sensationalism. Media outlets should work to present scientific discoveries and potential encounters with nuance and accuracy.

Preparing for Unpredictability:

The nature of an alien encounter is inherently unpredictable. Societies must be prepared for a wide range of potential scenarios, whether it involves the discovery of simple microbial life or the possibility of advanced extraterrestrial civilizations. Flexibility and adaptability are essential traits for navigating the unknown.

Education and Outreach:

Education and public outreach programs are essential to inform and engage the public about the search for extraterrestrial life. By fostering interest and understanding in these topics, we encourage critical thinking and promote informed discussions about the potential impact of an encounter.

Avoiding Panic and Hysteria:

While the prospect of alien contact is awe-inspiring, there is also a risk of generating unwarranted panic or fear. Responsible communication and education can help manage public reactions and ensure that any contact is met with a rational and composed response.

Preparing for Interstellar Diplomacy:

In the event of contact with an advanced extraterrestrial civilization, humanity must be prepared for interstellar diplomacy. Establishing protocols and guidelines for communication, collaboration, and potential conflict resolution will be crucial for navigating interstellar relations.

In conclusion, humanity's readiness for encountering extraterrestrial life extends far beyond scientific and technological preparedness. It requires ethical considerations, cultural sensitivity, responsible media reporting, and public engagement. As we continue to explore the cosmos and search for signs of life beyond Earth, it is essential to approach the possibility of contact with a thoughtful and holistic approach, fostering a united and informed response from humanity as a whole.

Chapter 11: Epilogue: Embracing the Unknown

The exploration of space, the search for extraterrestrial life

As we reach the end of our cosmic journey, we find ourselves standing at the threshold of the unknown, gazing out into the vast expanse of the universe. The exploration of space, the search for extraterrestrial life, and the quest for alien contact have been journeys that have ignited our curiosity, challenged our understanding, and inspired our imagination. In this epilogue, we reflect on the significance of our cosmic pursuits and the profound lessons they teach us about embracing the unknown.

Humility in the Face of the Cosmos

Our explorations of the universe have revealed the sheer scale of creation. The cosmos humbles us, reminding us of our place in the grand tapestry of existence. As we observe distant galaxies, ponder the mysteries of black holes, and search for life beyond Earth, we come to realize that our planet is but a tiny speck in the vastness of space. This humility encourages us to approach the unknown with an open mind and a willingness to learn.

The Beauty of Curiosity and Wonder

Curiosity is the spark that has driven us to reach for the stars. Wondering about the cosmos and its mysteries has fueled the human spirit for millennia. Our innate curiosity pushes the boundaries of what we know and drives us to seek answers to the most profound questions about our origins, our place in the universe, and the possibility of other intelligent beings. Embracing the unknown becomes an adventure, a journey of discovery that fuels our collective imagination.

Embracing Uncertainty as a Catalyst for Growth

The unknown can be a source of trepidation, but it is also a catalyst for growth and progress. In our pursuit of space exploration and the search for alien life, we face uncertainty at every turn. Yet, it is precisely in the face of uncertainty that we have made our greatest scientific and technological advancements. Embracing the unknown with courage and determination drives us forward as a species, propelling us to overcome challenges and broaden our horizons.

A Shared Human Endeavor

Our cosmic journey is not limited to scientists, astronauts, or space agencies alone. It is a shared human endeavor that transcends borders, cultures, and backgrounds. The wonder of the cosmos unites us as a species, reminding us that we are all travelers on this pale blue dot we call Earth. Embracing the unknown together, we find strength in our shared aspirations, dreams, and hope for a brighter future.

Gratitude for the Beauty of Creation

The universe presents us with breathtaking beauty—glittering galaxies, swirling nebulae, and distant stars that shine like celestial diamonds. Our explorations have gifted us the opportunity to witness this splendor, instilling a profound sense of gratitude for the beauty of creation. Embracing the unknown with gratitude allows us to savor each discovery as a gift from the cosmos.

A Journey of Perpetual Wonder

As we conclude this journey through the cosmos and our search for extraterrestrial life, we realize that the quest for knowledge and understanding is an unending voyage of perpetual wonder. The universe is an infinite canvas of possibilities, inviting us to continue exploring, discovering, and imagining what lies beyond the horizon.

Embracing the unknown is not a destination but a way of life—an invitation to look up at the stars, ponder our place in the cosmos, and marvel at the wonders of creation. The universe beckons us to embrace the unknown with open hearts and curious minds, for it is in the mystery that we find the true essence of our humanity. As we venture forth, we carry the spirit of exploration, curiosity, and wonder, ever eager to explore new frontiers and discover the unfathomable mysteries that await us in the great cosmic expanse.

Reflecting on the Wonders of the Universe and the Search for Alien Life

As we reflect on the wonders of the universe and our quest to find signs of alien life, we are reminded of the sheer majesty and complexity of creation. The universe, with its galaxies, stars, planets, and mysteries, captivates our imagination and evokes a sense of awe and curiosity. In this contemplation, we discover profound insights about ourselves and our place in the cosmos.

Embracing Our Place in the Cosmos

The search for alien life reminds us that we are not isolated beings but part of a vast interconnected web of life that spans the cosmos. Every atom in our bodies was forged in the heart of a star, and every element on Earth originates from the crucibles of distant celestial bodies. Embracing our cosmic origins fosters a sense of interconnectedness and unity with the universe.

The Beauty of Diversity

The universe is a tapestry of diversity, with a kaleidoscope of planets, stars, and galaxies. Our search for alien life reflects our fascination with the possibility of encountering life forms vastly different from our own. The wonders of the universe teach us to appreciate the beauty of diversity, both here on Earth and potentially beyond.

Humility in the Face of the Unknown

As we explore the cosmos and search for alien life, we encounter mysteries that challenge our understanding. The more we learn, the more we realize how much we don't know. Embracing the unknown with humility allows us to remain open to new discoveries and encourages a lifelong pursuit of knowledge.

The Persistence of Human Curiosity

Throughout history, human curiosity has driven us to explore and push the boundaries of our knowledge. The search for alien life is a testament to the enduring spirit of curiosity and the belief that there is much more to the cosmos than what we currently comprehend. It is this persistence of curiosity that propels us forward, sparking new questions and igniting our passion for discovery.

Wondering About Our Place in the Cosmic Story

The universe has been evolving for billions of years, and our brief existence is just a fleeting chapter in its cosmic story. Contemplating the vastness of time and space leads us to ponder our place in this grand narrative. The search for alien life allows us to glimpse into the possibility of other civilizations and their stories, reflecting on how our own story intertwines with the broader cosmic tapestry.

Gratitude for the Gift of Consciousness

As we explore the cosmos and contemplate the search for alien life, we recognize the remarkable gift of consciousness that allows us to ponder these questions. We marvel at our capacity to observe, question, and explore the universe. Embracing this gift with gratitude deepens our connection to the cosmos and encourages us to be stewards of life and the planet we call home.

The Ongoing Journey

The wonders of the universe and the search for alien life are not finite quests with conclusive endpoints. Instead, they are ongoing journeys of exploration and understanding. The cosmos will continue to reveal its secrets, and the search for alien life will persist as long as we exist as a curious and adventurous species.

In conclusion, reflecting on the wonders of the universe and our search for alien life invites us to embrace the marvels of creation, our place in the cosmic story, and the gift of human consciousness. It fosters a deep sense of wonder, humility, and gratitude for the mysteries that await us in the cosmos. As we continue our journey of exploration, we are reminded that the pursuit of knowledge and understanding is not just about finding answers but about embracing the beauty of the unknown and the infinite possibilities that lie ahead.

The Importance of Continued Exploration and Curiosity

Exploration and curiosity are two of humanity's most potent driving forces, propelling us to seek knowledge, expand our horizons, and venture into the unknown. Throughout history, these qualities have led to extraordinary discoveries and advancements that have shaped our understanding of the world and the universe. In this discussion, we explore the importance of continued exploration and curiosity and the profound impact they have on society, science, and the human spirit.

Advancing Scientific Knowledge:

Curiosity fuels scientific inquiry, encouraging researchers and explorers to ask questions, challenge assumptions, and seek evidence-based answers. The spirit of exploration drives us to push the boundaries of human understanding, leading to groundbreaking discoveries in various fields, from physics and astronomy to biology and medicine. Continued exploration allows us to expand the frontiers of knowledge, unravel mysteries, and make leaps in technological innovations that benefit humanity.

Fostering Innovation and Progress:

Exploration and curiosity are the catalysts for innovation and progress. The pursuit of new ideas and the desire to solve complex problems spark the creation of novel technologies and solutions. Throughout history, explorations of uncharted territories have led to remarkable advancements in navigation, transportation, communication, and numerous other areas that have transformed the way we live and interact with the world.

Inspiring Future Generations:

The spirit of exploration and curiosity inspires the next generation of scientists, engineers, artists, and leaders. The fascination with the wonders of the universe and the desire to understand the natural world motivate young minds to pursue education, engage in research, and dream big. By nurturing curiosity, we ensure a continuous cycle of innovation and progress that benefits society for generations to come.

Fostering a Global Perspective:

Continued exploration fosters a global perspective by encouraging collaboration, cooperation, and cultural exchange. International missions, space explorations, and scientific research projects often involve collaboration among scientists, engineers, and professionals from different countries. This shared pursuit of knowledge transcends political boundaries and promotes a sense of interconnectedness among nations.

Addressing Global Challenges:

Curiosity-driven exploration plays a crucial role in addressing global challenges, such as climate change, environmental degradation, and public health crises. By exploring and understanding the complexities of our planet and the cosmos, we gain insights into the interconnected systems that sustain life. This knowledge enables us to develop sustainable solutions and make informed decisions to address pressing global issues.

Nurturing the Human Spirit:

Exploration and curiosity nourish the human spirit by instilling a sense of wonder, awe, and humility. The unknown beckons us to embark on journeys of discovery, fueling our imagination and desire to explore not just the physical world but also the depths of our own minds and consciousness. In this introspective exploration, we find meaning and purpose, fostering personal growth and fulfillment.

Embracing Uncertainty:

Continued exploration and curiosity teach us to embrace uncertainty as an opportunity for growth and learning. The willingness to venture into the unknown, acknowledge gaps in knowledge, and seek answers with an open mind leads to resilience and adaptability. By embracing uncertainty, we become better equipped to navigate the challenges of an ever-changing world.

In conclusion, continued exploration and curiosity are indispensable elements of the human experience. They drive us to seek knowledge, push the boundaries of understanding, and shape the course of human history. By nurturing these qualities in ourselves and future generations, we ensure that the spirit of exploration remains alive, inspiring us to uncover the wonders of the universe, address global challenges, and embark on transformative journeys that enrich our lives and deepen our connection with the cosmos.

Embracing the Unknown with Awe and Humility

In a world filled with scientific advancements and technological marvels, it is easy to believe that we have unraveled all the mysteries of the universe. However, as we peer into the vastness of space and explore the intricacies of life, we are humbled by the realization that the unknown remains an integral part of our existence. Embracing the unknown with awe and humility is a profound way of engaging with the mysteries that surround us, inspiring a deeper appreciation for the wonders of life and the cosmos.

Awe: The Gateway to Wonder

Awe is an emotion that arises when we encounter something vast, transcendent, or incomprehensible. It is the feeling that washes over us when we gaze at the starry night sky, witness a majestic natural phenomenon, or ponder the complexity of life's origins. Embracing the unknown with awe opens our hearts and minds to the infinite possibilities that lie beyond our current understanding. It reminds us of the limitations of our knowledge and fuels the flame of curiosity, beckoning us to explore the uncharted territories of science, philosophy, and spirituality.

Humility: Recognizing Our Place in the Cosmos

Embracing the unknown with humility involves recognizing that, despite our intellectual achievements, we are but a small part of the cosmic tapestry. Humility urges us to acknowledge that there is much we do not know and that our understanding of the universe is only a fraction of its true depth. As we contemplate the grandeur of the cosmos and the vastness of time, we come to understand that our existence is fleeting, and our species is but a brief chapter in the cosmic narrative. This humility fosters a deep reverence for the mysteries of life and a sense of gratitude for the opportunity to be part of such an awe-inspiring cosmos.

The Intersection of Awe and Humility

The intersection of awe and humility lies at the heart of our relationship with the unknown. Awe opens our minds to the beauty and complexity of existence, while humility keeps us grounded, reminding us that our pursuit of knowledge is an ongoing journey with no end in sight. Together, these emotions cultivate a sense of wonder that infuses our lives with meaning and purpose.

Embracing Uncertainty as a Source of Growth

Embracing the unknown with awe and humility allows us to view uncertainty as an opportunity for growth and self-discovery. Rather than fearing the unknown, we learn to navigate it with grace and courage, drawing inspiration from the vastness of the cosmos and the resilience of life. In moments of uncertainty, we find the courage to explore new paths, challenge conventional wisdom, and forge ahead in the face of adversity.

The Power of Wonder in Human Experience

Wonder is the soul's response to the mysteries of life and the cosmos. It fuels our passion for discovery, drives us to seek deeper connections, and fosters a sense of interconnectedness with all living beings. Embracing the unknown with awe and humility nourishes this wonder, allowing us to marvel at the intricate dance of existence and the potential for infinite possibilities.

In conclusion, embracing the unknown with awe and humility is a transformative approach to engaging with the mysteries of life and the universe. It invites us to see beyond the limits of our current knowledge and embrace the beauty of uncertainty. By cultivating a spirit of wonder and recognizing our place in the grand cosmic scheme, we embark on a journey of exploration that enriches our lives, deepens our understanding, and connects us to the awe-inspiring mysteries that await us in the vastness of the unknown.

Appendix: Resources and References

The exploration of the universe and the search for alien life are fields that draw upon a vast array of scientific research, literature, and media. Here are some valuable resources and references that can further enrich your understanding of these captivating subjects:

Books:

"Cosmos" by Carl Sagan: A classic exploration of the cosmos, weaving science, history, and philosophy to provide a grand perspective of the universe.

"Pale Blue Dot: A Vision of the Human Future in Space" by Carl Sagan: A thought-provoking book that reflects on humanity's place in the universe and the potential for space exploration.

"Astrophysics for People in a Hurry" by Neil deGrasse Tyson: A concise and accessible overview of astrophysics that covers a wide range of cosmic topics.

"The Search for Extraterrestrial Life: A Philosophical Inquiry" by Milan M. Ćirković: This book delves into the philosophical implications of the search for alien life and the possible consequences of contact.

Websites:

NASA's Exoplanet Exploration: An official NASA website dedicated to exoplanet exploration, providing up-to-date information and data on exoplanets and their potential habitability.

SETI Institute: The Search for Extraterrestrial Intelligence (SETI) Institute conducts research and educational programs focused on the quest for alien civilizations.

European Space Agency (ESA): ESA's website offers information on space missions, including those related to exoplanet studies and cosmic exploration.

Astrobiology Web: A comprehensive resource for astrobiology, covering various topics related to the search for life beyond Earth.

Scientific Papers and Journals:

The Astrophysical Journal (ApJ): A prominent scientific journal publishing research in astrophysics, including exoplanet discoveries and studies.

Astrobiology Journal: A peer-reviewed journal that covers the interdisciplinary field of astrobiology, exploring the origin and potential for life in the universe.

Nature Astronomy: A leading scientific journal focusing on astronomy and astrophysics, featuring cutting-edge research in the field.

Documentaries and Films:

"Cosmos: A Spacetime Odyssey": This documentary series, hosted by Neil deGrasse Tyson, explores the wonders of the universe and humanity's quest for knowledge.

"The Farthest: Voyager in Space": A documentary that chronicles the incredible journey of the Voyager spacecraft and its encounters with other planets.

"The Search for Life in Space": A documentary that delves into the science and technology behind the search for extraterrestrial life.

These resources provide a glimpse into the vast realm of knowledge and wonder that surrounds the exploration of the universe and the search for alien life. They offer a starting point for those curious to delve deeper into these captivating subjects, inspiring a lifelong journey of discovery and appreciation for the mysteries that await us in the cosmos.

Recommended Books, Articles, and Documentaries for Further Exploration

If you find yourself intrigued by the wonders of the universe and the search for alien life, the following list of recommended books, articles, and documentaries will provide you with a rich and diverse exploration of these captivating subjects:

Books:

"Cosmos" by Carl Sagan: A timeless classic that takes readers on a journey through space and time, exploring the wonders of the universe and our place within it.

"Astrophysics for People in a Hurry" by Neil deGrasse Tyson: A concise and engaging book that offers a quick yet informative overview of astrophysics and cosmology.

"Exoplanets: Diamond Worlds, Super Earths, Pulsar Planets, and the New Search for Life Beyond Our Solar System" by Michael Summers and James Trefil: An exploration of the exciting field of exoplanet research and the potential for life on other worlds.

"The Copernicus Complex: Our Cosmic Significance in a Universe of Planets and Probabilities" by Caleb Scharf: A thought-provoking examination of humanity's place in the cosmos and the implications of the discovery of exoplanets.

Articles and Scientific Papers:

"The Drake Equation": Explore the famous equation that estimates the number of potential extraterrestrial civilizations in our galaxy.

"The Fermi Paradox": Investigate the paradox that questions why, despite the vast number of potential civilizations, we have not yet encountered any extraterrestrial intelligence.

"Astrobiology: A Brief Introduction": Delve into the field of astrobiology, which seeks to understand the origin, evolution, and distribution of life in the universe.

"Water Worlds: The Search for Habitable Exoplanets":
Learn about the search for exoplanets with liquid water—a key ingredient for life as we know it.

Documentaries and Films:

"The Universe": A comprehensive documentary series that explores various aspects of the cosmos, including black holes, galaxies, and the potential for alien life.
"The Search for Life: The Drake Equation": This documentary investigates the implications of the Drake Equation and the possibility of finding intelligent life in the universe.
"Life Beyond Earth": A National Geographic documentary that examines the latest scientific discoveries related to the search for extraterrestrial life.
"The Farthest: Voyager in Space": A captivating documentary that chronicles the journey of the Voyager spacecraft, humanity's farthest-reaching mission into space.

Websites and Organizations:

NASA's Exoplanet Exploration: Access up-to-date information on exoplanets and NASA's efforts to explore these distant worlds.
SETI Institute: Learn about the search for extraterrestrial intelligence and the latest developments in the quest for alien signals.
European Southern Observatory (ESO): Explore the latest astronomical discoveries and observatory missions in Europe.
Astrobiology Web: A comprehensive website that covers various topics related to astrobiology and the search for life beyond Earth.

By engaging with these resources, you can deepen your understanding of the universe's wonders and the ongoing pursuit of knowledge about extraterrestrial life. Whether you prefer to read in-depth scientific literature, explore thought-provoking articles, or watch captivating documentaries, these recommendations offer a diverse range of content to satisfy your curiosity and spark new questions about the cosmos. Happy exploring!

Online Resources for Space Enthusiasts and UFO Researchers

For space enthusiasts and UFO researchers alike, the internet offers a wealth of information, data, and community engagement. Here are some valuable online resources to quench your thirst for cosmic knowledge and UFO-related investigations:

Space Exploration and Astronomy:

NASA Website: The official website of the National Aeronautics and Space Administration (NASA) provides a vast repository of space-related information, mission updates, images, and educational resources. Visit: www.nasa.gov

European Space Agency (ESA): ESA's website offers news, updates, and insights into space missions and astronomical research conducted by the European space agency. Visit: www.esa.int

Sky & Telescope: This website is a go-to resource for astronomy enthusiasts, featuring news, articles, celestial events, and tips for stargazing. Visit: skyandtelescope.org

Space.com: A popular space news website covering the latest discoveries, space missions, and cosmic events. Visit: www.space.com

UFO and Extraterrestrial Research:

MUFON (Mutual UFO Network): A nonprofit organization dedicated to investigating UFO sightings and promoting the scientific study of UFOs. Their website offers reports, case studies, and research articles. Visit: www.mufon.com

UFO Stalker: A user-generated UFO sighting database that maps UFO reports worldwide. Visitors can explore and submit their sightings. Visit: www.ufostalker.com

The Black Vault: This website, run by UFO researcher John Greenewald Jr., hosts an extensive collection of declassified UFO documents, government files, and other related materials. Visit: www.theblackvault.com

UFO Sightings Daily: A blog by Scott C. Waring, which features UFO sightings, alleged alien encounters, and discussions on extraterrestrial phenomena. Visit: www.ufosightingsdaily.com

Scientific Research Journals:

Journal of UFO Studies: A peer-reviewed journal that publishes research on UFO sightings, encounters, and related topics. Visit: www.jufos.org

International Journal of Astrobiology: This journal covers a broad range of astrobiology topics, including the search for extraterrestrial life. Visit: www.cambridge.org/ijb

Astrobiology Journal: A leading scientific journal in astrobiology, focusing on the origin, evolution, and distribution of life in the universe. Visit: www.liebertpub.com/ast

UFO and Alien Encounter Podcasts:

The Black Vault Radio with John Greenewald Jr.: A podcast that explores UFOs, government secrets, and the unexplained with notable guests. Available on major podcast platforms.

Mysterious Universe: A long-running podcast that covers a wide range of paranormal and UFO-related topics, including sightings, encounters, and conspiracies. Available on major podcast platforms.

UFO Chronicles Podcast: A podcast that delves into UFO sightings, encounters, and interviews with witnesses and researchers. Available on major podcast platforms.

Online Forums and Communities:

Reddit - r/UFOs: An active community for UFO enthusiasts to share sightings, discuss theories, and engage in open dialogue about UFO-related topics. Visit: www.reddit.com/r/UFOs

AboveTopSecret: A popular forum for discussions on UFOs, conspiracy theories, and alternative topics. Visit: www.abovetopsecret.com

UFO Seekers: An online community of UFO enthusiasts sharing stories, research, and experiences related to UFOs. Visit: www.ufoseekers.com

Remember to critically assess the information you find online and consider multiple sources to form well-rounded perspectives. These resources can serve as valuable starting points for space exploration and UFO research, providing a platform to delve deeper into the fascinating realms of the cosmos and the unexplained. Happy exploring and researching!